AF595335

Environmental Hydraulics, Turbulence and Sediment Transport

Environmental Hydraulics, Turbulence and Sediment Transport

Editor

Jaan H. Pu

MDPI • Basel • Beijing • Wuhan • Barcelona • Belgrade • Manchester • Tokyo • Cluj • Tianjin

Editor
Jaan H. Pu
Faculty of Engineering and Informatics,
University of Bradford
UK

Editorial Office
MDPI
St. Alban-Anlage 66
4052 Basel, Switzerland

This is a reprint of articles from the Special Issue published online in the open access journal *Fluids* (ISSN 2311-5521) (available at: https://www.mdpi.com/journal/fluids/special_issues/hydraulics_bedforming).

For citation purposes, cite each article independently as indicated on the article page online and as indicated below:

LastName, A.A.; LastName, B.B.; LastName, C.C. Article Title. *Journal Name* **Year**, *Volume Number*, Page Range.

ISBN 978-3-0365-3242-4 (Hbk)
ISBN 978-3-0365-3243-1 (PDF)

Cover image courtesy of Aiguli Talpakova.

Contents

About the Editor

Jaan H. Pu Dr Pu received both his BEng (1st Class Honours) and PhD degrees from the University of Bradford in 2003 and 2008, respectively. Since 2008, he has taken up several research and faculty positions at various universities/institutions around the world. He was appointed as a Lecturer in Civil Engineering at the University of Bradford in 2014; as Senior Lecturer in 2017; and as Associate Professor in 2020. Dr Pu's research concentrates on numerical and laboratory approaches to representing various water engineering applications which, include natural compound riverine flow, sediment transport, scouring, water quality, and vegetated flow. His research outputs have led to several high-quality journal articles (50+), conference proceedings (10+), edited books (2), and book chapters (2). He is currently supervising four PhD students at Bradford (three as their principal supervisor) who are investigating river hydrodynamics and sediment transport applications. He has also been appointed as the Leading Guest Editor of Special Issues, such as The Urban Fluvial and Hydro-Environment System (Frontiers in Environmental Science), Environmental Sediment Transport: Methods and Applications (MDPI Fluids), and Environmental Hydraulics, Turbulence and Sediment Transport (MDPI Fluids); and as the Guest Editor of a Special Issue on Advances in Modelling and Prediction on the Impact of Human Activities and Extreme Events on Environments (MDPI Water). He is currently a Visiting Scientist at Tsinghua University and Nanyang Technological University. He has been involved as a reviewer for several internationally well-reputed journals.

Preface to "Environmental Hydraulics, Turbulence and Sediment Transport"

In the research on environmental hydraulics, its turbulence and sediment transport, constant challenges have been faced. The complexity of hydraulic impacts towards sediment morphology and turbulent flow properties makes research in this area a difficult task. However, due to pressure from climate change and the mounting issue of pollution, environmental flow studies are more crucial than ever.

Bedforming within rivers is a complex process that can be influenced by the hydraulics, vegetated field, and various suspended and bedload transports. Changes in flow conditions due to rain and flood can further complicate a hydraulic system. To date, the turbulence, morphologic, and bedforming characteristics of natural environmental flows are still not well understood.

This book aims to bring together a collection of state-of-the-art research and technologies to form a useful guide for the related research and engineering communities. It is useful for authorities and researchers interested in environmental and civil engineering studies, as well as for river and water engineers to understand the current state-of-the-art practices in environmental flow modelling, measurement and management. It is also a good resource for research, post-, or undergraduate students who wish to know about the most up-to-date knowledge in this field.

Jaan H. Pu
Editor

 MDPI

Editorial

Environmental Hydraulics, Turbulence and Sediment Transport

Jaan H. Pu

Faculty of Engineering and Informatics, University of Bradford, Bradford DB7 1DP, UK; j.h.pu1@bradford.ac.uk; Tel.: +44-01274234556

Citation: Pu, J.H. Environmental Hydraulics, Turbulence and Sediment Transport. *Fluids* **2022**, *7*, 48. https://doi.org/10.3390/fluids7020048

Received: 19 January 2022
Accepted: 21 January 2022
Published: 23 January 2022

Publisher's Note: MDPI stays neutral with regard to jurisdictional claims in published maps and institutional affiliations.

Within the environmental flows, i.e., river and canal networks or irrigation systems, hydraulic properties for water flow are usually impacted by eddies and sediment transport. Thus, the accuracy in predicting natural environmental flow behaviour is heftily related to the precise knowledge on its morphological characteristics under naturally occurring turbulence circumstances. The deposition and erosion under sediment-laden flow also play a crucial role in how the natural flows and watercourses are supposed to be managed to alleviate and avoid disastrous pollution transport.

In a previous Special Issue (SI) of *Fluids* entitled "Environmental Sediment Transport: Methods and Applications", the sediment transport phenomenon has been explored in the applications of natural bedform (Pu [1]) and deposition within rainwater (John et al. [2]). As a continuation, the leading guest editor for that SI decided to launch a new volume to further advance and expand on the discussion topics. This SI, which contains a collection of seven papers, gathers recent advances in the fields of environmental hydraulics, turbulence and sediment transport from various international authors/researchers to tackle the above-mentioned important and pressing environmental issues.

Owing to the fact that suspended solid transport is among the important factors to alleviate pollutant propagation, its mathematical modelling has been intensely investigated in recent literature studies. In the review work by Wallwork et al. [3], the previous research on mathematical modelling of suspended sediment transport has been summarised and analysed. Besides numerical models, the suggested analytical approaches in the literature to represent the suspended load have also been reviewed. Wallwork et al.'s study increases the potential for researchers to learn from previous experiences and studies in order to move towards the goal for accurate sediment transport modelling.

Sediment-laden condition is regarded as one of the key mechanisms to the formation of rough bedform in natural flow (Pu [1]). To add to the complexity, any obstruction like wood-log or boulder, in particular those obstructions across channel width, can attach with the bedform during its formation to cause floods. To study the impact of the horizontal obstruction in rough bed flow, Devi et al. [4,5] have run a series of experiments by bed-mounted cylinder to investigate their flow velocity [4] and turbulence [5]. Using an acoustic Doppler velocimetry (ADV) measurement approach, Devi et al. [4] concentrated on comparison between flow properties impacted by sand and gravel-formed beds, while Devi et al. [5] considered the turbulence of flow above sand bed by high-order bursting analysis. The findings by [4,5] are crucial for consolidating the knowledge of turbulence field induced by the horizontally laying cylindrical obstacle and predicting its actual flow characteristic over a rough bed.

Tailing dam has been studied by Satyanaga et al. [6], where its sediment tailings have been investigated by incorporating the unsaturated solid mechanic approach. In their study, the soil consolidation process was illustrated to observe the stability of sediment within the dam. The experiment together with numerical analysis proposed within Satyanaga et al.'s study is particularly crucial to understand and manage the operation of the tailing dam. To better understand the dynamics of dredging and its impact on sediment transport within the natural river/watercourse environment, Zikra et al. [7] has studied a case at Nagan Raya Port of Indonesia. In their work, the numerical modelling and field study have been co-utilised, where the results showed a severe impact of dredging towards negatively promoting the sediment transport surrounding the port. The study by Zikra et al. [7] presented useful guidelines not only to the local Nagan Raya Port, but also to portA with similar natural environments.

In supporting structures above the water, various stability research has been performed for many common water-based structures. In the investigation by Bento et al. [8], the flow field around the bridge pier's scour has been studied. The oblong bridge piers have been concentrated in their study, where the proposed experimental results can explore the scour development pattern and hence is useful for the stability study of the bridge's structure. Besides rigid structures like bridges, the floating structures with high degrees of flexibility are also popularly used in water-based projects, e.g, as petroleum drilling platforms. Cui et al. [9] studied the hydrodynamic moment of such flexible structure using 121 sets of detailed experiments. Their results can shed the light on the relative industry in terms of the structural stability and durability subject to constant hydrodynamic loading.

Finally, it is also important to recognise the effort by anonymous reviewers to the above-mentioned articles. Without their contributions, this Special Issue would not be possible.

Funding: This research received no external funding.

Acknowledgments: Throughout the editing of the whole Special Issue and the writing of this Editorial article, the author has been greatly inspired and supported by his family members, Aigul, Jasmin and Jeanette. Without their tireless supports, all these will not be possible.

Conflicts of Interest: The author declares no conflict of interest.

References

1. Pu, J.H. Velocity Profile and Turbulence Structure Measurement Corrections for Sediment Transport-Induced Water-Worked Bed. *Fluids* **2021**, *6*, 86. [CrossRef]
2. John, C.K.; Pu, J.H.; Pandey, M.; Hanmaiahgari, P.R. Sediment Deposition within Rainwater: Case Study Comparison of Four Different Sites in Ikorodu, Nigeria. *Fluids* **2021**, *6*, 124. [CrossRef]
3. Wallwork, J.T.; Pu, J.H.; Kundu, S.; Hanmaiahgari, P.R.; Pandey, M.; Satyanaga, A.; Khan, M.A.; Wood, A. Review of Suspended Sediment Transport Mathematical Modelling Studies. *Fluids* **2022**, *7*, 23. [CrossRef]
4. Devi, K.; Hanmaiahgari, P.R.; Balachandar, R.; Pu, J.H. A Comparative Study between Sand- and Gravel-Bed Open Channel Flows in the Wake Region of a Bed-Mounted Horizontal Cylinder. *Fluids* **2021**, *6*, 239. [CrossRef]
5. Devi, K.; Hanmaiahgari, P.R.; Balachandar, R.; Pu, J.H. Self-Preservation of Turbulence Statistics in the Wall-Wake Flow of a Bed-Mounted Horizontal Pipe. *Fluids* **2021**, *6*, 453. [CrossRef]
6. Satyanaga, A.; Wijaya, M.; Zhai, Q.; Moon, S.-W.; Pu, J.H.; Kim, J.R. Stability and Consolidation of Sediment Tailings Incorporating Unsaturated Soil Mechanics. *Fluids* **2021**, *6*, 423. [CrossRef]
7. Zikra, M.; Salsabila, S.; Sambodho, K. Toward a Better Understanding of Sediment Dynamics as a Basis for Maintenance Dredging in Nagan Raya Port, Indonesia. *Fluids* **2021**, *6*, 397. [CrossRef]
8. Bento, A.M.; Viseu, T.; Pêgo, J.P.; Couto, L. Experimental Characterization of the Flow Field around Oblong Bridge Piers. *Fluids* **2021**, *6*, 370. [CrossRef]
9. Cui, Z.; Pan, S.-Y.; Chen, Y.-J. The Hydrodynamic Moment of a Floating Structure in Finite Flowing Water. *Fluids* **2021**, *6*, 307. [CrossRef]

MDPI

Review

Review of Suspended Sediment Transport Mathematical Modelling Studies

Joseph T. Wallwork [1], Jaan H. Pu [2,*], Snehasis Kundu [3], Prashanth R. Hanmaiahgari [4], Manish Pandey [5], Alfrendo Satyanaga [6], Md. Amir Khan [7] and Alastair Wood [2]

1 Mott MacDonald Bentley, Skipton BD23 2QR, UK; theowallwork@gmail.com
2 Faculty of Engineering and Informatics, University of Bradford, Bradford DB7 1DP, UK; a.s.wood@bradford.ac.uk
3 Department of Mathematics, National Institute of Technology Jamshedpur, Jharkhand 831014, India; snehasis18386@gmail.com
4 Department of Civil Engineering, Indian Institute of Technology Kharagpur, West Bengal 721302, India; hpr@civil.iitkgp.ac.in
5 Department of Civil Engineering, Water Resources and Environmental Division, National Institute of Technology Warangal, Warangal 506004, India; manishpandey3aug@gmail.com
6 Department of Civil and Environmental Engineering, Nazarbayev University, Kabanbay Batyr Ave., Nur-Sultan 010000, Kazakhstan; alfrendo.satyanaga@nu.edu.kz
7 Galgotias College of Engineering and Technology, Greater Noida, Uttar Pradesh 201310, India; amirmdamu@gmail.com
* Correspondence: j.h.pu1@bradford.ac.uk

Citation: Wallwork, J.T.; Pu, J.H.; Kundu, S.; Hanmaiahgari, P.R.; Pandey, M.; Satyanaga, A.; Khan, M.A.; Wood, A. Review of Suspended Sediment Transport Mathematical Modelling Studies. *Fluids* **2022**, *7*, 23. https://doi.org/10.3390/fluids7010023

Academic Editor: Mehrdad Massoudi

Received: 6 November 2021
Accepted: 23 December 2021
Published: 3 January 2022

Abstract: This paper reviews existing studies relating to the assessment of sediment concentration profiles within various flow conditions due to their importance in representing pollutant propagation. The effects of sediment particle size, flow depth, and velocity were considered, as well as the eddy viscosity and Rouse number influence on the drag of the particle. It is also widely considered that there is a minimum threshold velocity required to increase sediment concentration within a flow above the washload. The bursting effect has also been investigated within this review, in which it presents the mechanism for sediment to be entrained within the flow at low average velocities. A review of the existing state-of-the-art literature has shown there are many variables to consider, i.e., particle density, flow velocity, and turbulence, when assessing the suspended sediment characteristics within flow; this outcome further evidences the complexity of suspended sediment transport modelling.

Keywords: suspended sediment concentration; dilute-hyper concentration; Rouse number; velocity lag; bursting phenomena

1. Introduction

Sediment suspension describes the solid particles that have been lifted into the water column from the channel bed. It is beneficial to have a thorough understanding of sediment transport and thus two-phase flow to allow for its effective mathematical modelling, since it is a regular occurrence within large water bodies in the natural environment. This study looks to review a range of existing literature and compare proven evidence to provide a sound understanding of solid-fluid interactions within two-phase flow conditions.

When there is sufficient lift force for sediment particles to overcome the friction between them, the turbulent upward flux will generate sediment suspension. Generally, it is accepted that the mean concentration decreases with height above the bed, as shown in Figure 1 and described by [1].

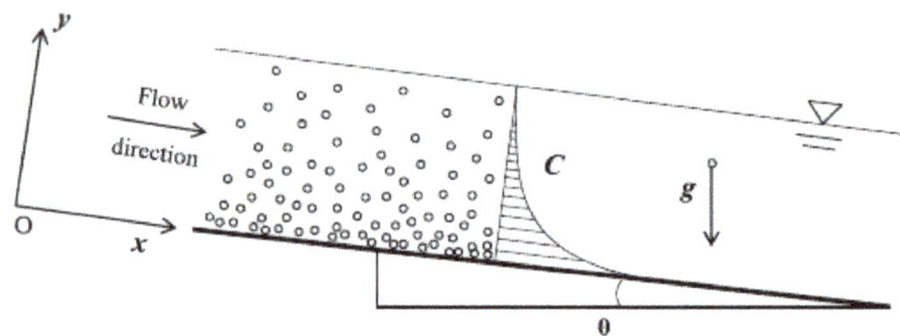

Figure 1. Sediment Concentration Profile. Reprinted with permission from [1]. 2021 Hanmaiahgari, P.R.

Two-phase flow is the combination of two states of matter flowing together; in this case, solid and liquid. During two-phase sediment transport flow, there are complex interactions between the solid and fluid phases. This complexity has been presented in various studies, such as on natural channel flow [2,3], flow with natural bedform [4,5], and sediment transport modelling [6,7]. These interactions can be difficult to model mathematically due to the large number of variables present within open channel flow and the chaotic nature of fluid dynamics. Early methods of modelling sediment concentration [8–12] resolved some of these difficulties by ignoring forces acting on the sediment particle. However, applying these assumptions limits the model accuracy to certain sediment transport conditions, e.g., light particle transport. To create a more holistic formula, recent research has incorporated other forces acting on the particle phase [13–15]. This has resulted in some diverse formulations for predicting the sediment profile.

A popular consensus when assessing two-phase flow is to consider a two-dimensional (2D) observation, i.e., look at how concentration varies with depth and along the streamwise space. The considered parameters to assess such flow often include fluid velocity, particle diameter, Rouse number, and mean concentration [16–19]. Other research has incorporated the velocity fluctuations due to drift-flux and vortices considering eddy viscosity [19–21].

Most research into sedimentology has associated suspension with a time-averaged shear stress at the bed, whereas further studies have shown that sediment suspension can occur due to instantaneous velocity fluctuations in the vertical direction at the channel bed [22–24]. A concept to describe this is that the sediment near the bed experiences the highest concentration gradient and thus a high-pressure gradient, which forces the particle into suspension. This corresponds to the concentration profile shown in Figure 1. However, recent observations have also further revealed alternative sediment concentration profiles where concentration peaks at a height above the channel bed region. This has been shown to be largely dependent on the various concentration of sediment entrained within a flow from dilute to dense conditions [18].

2. Literature Review

Several methods can be used to model sediment concentration, which the diffusion and kinetic theories are among the most common. Diffusion theory, which is recognized as one of the simplest methods for modelling sediment concentration, has been used by many scientists to describe the solid phase with reasonable accuracy [11,21]. On the other hand, the kinetic theory is widely regarded as the best approach for concentration distribution modelling as it includes the response of both the solid and liquid phases as well as the two-phase interactions [13]. Extensive research has been undertaken to understand and model these two-phase interactions and their effects on the sediment profile [9,15,19,25,26], where they will be explored in the coming sub-sections.

2.1. Reynolds Number Approach

Drag is a mechanical force that arises due to the interaction between a body moving through a fluid and the induced resultant opposing frictional forces, where it can be represented by the following equation:

$$F_D = \frac{1}{2} C_D \rho_f A {v_r}^2 \tag{1}$$

where F_D denotes drag force, C_D is the drag coefficient, ρ_f is the density of fluid, and u_r is the relative velocity of the particle. Of the parameters contained within Equation (1), C_D is the most ambiguous and requires careful consideration. C_D can be defined as a constant [27]:

$$C_D = 0.44 \tag{2}$$

In multiple studies, constant C_D is shown to be inconsistent with reality. Therefore, a more robust approach was developed. Following further research [28–30], it is generally considered that C_D is proportional to Reynolds number (Re), and extensive studies have been performed to confirm this relationship.

It has been stated that for a perfectly smooth spherical particle, the drag coefficient shares an asymptotic relationship with Reynolds number for both laminar and turbulent flows [28]. Further study concluded that for rough particles, the relationship varies and is dependent on the flow regime. This is due to stresses and frictional forces arising from the boundary conditions. For example, Cheng [30] studied the settling velocity and effects of drag on experimental spherical particles. Within that study, a new equation for C_D was formed and compared with laboratory data. Figure 2 shows Equations (3)–(6) fitted against experimental data. The curve plotted in Figure 2 is subdivided into three sections. For laminar flow ($Re < 1$), C_D can be defined as:

$$C_D = \frac{24}{Re} \tag{3}$$

The flow becomes transitional at $1 \le Re < 100$, where C_D is defined as:

$$C_D = \frac{24}{Re}(1 + 0.27Re)^{0.43} \tag{4}$$

Finally, when the flow becomes turbulent ($Re \ge 100$), C_D can be represented by:

$$C_D = 0.47\left[1 - exp\left(-0.04Re^{0.38}\right)\right] \tag{5}$$

Equations (3)–(5) can then be combined to produce one explicit function for C_D, defined as [30]:

$$C_D = \frac{24}{Re}(1 + 0.27Re)^{0.43} + 0.47\left[1 - exp\left(-0.04Re^{0.38}\right)\right] \tag{6}$$

Alternatively, Equation (6) can be rewritten as:

$$C_D = \frac{24}{Re_p} + f(Re_p) \tag{7}$$

where f is a function of Re_p, in which Re_p is the particulate Reynolds number. The first term on the right-hand side of Equation (7) relates to laminar flow. The second term represents the drag force due to turbulence [13].

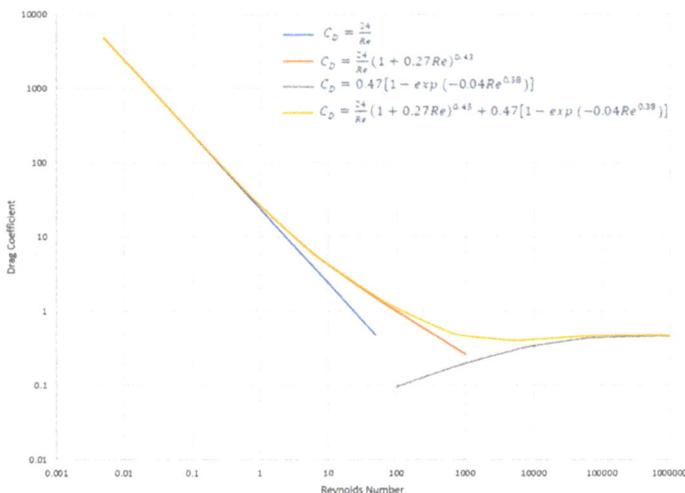

Figure 2. Coefficient of Drag and Reynolds Number Correlation (Adapted from Cheng [30]).

2.2. Velocity Lag Approach

The effects of drag force can also be assessed in terms of the velocity lag, which is the time differential for the momentum to transfer from the fluid phase to the solid phase [13,20,26]. Utilizing the fluid viscosity, Stokes' law studies the resistance to motion when an object is dropped in a fluid. When falling due to gravity, the object is acted upon by an equal and opposite reaction, as stated by Newton's third law. Therefore, when a particle is falling in laminar flow with no turbulent flux, the exerting forces are only those which arise due to fluid inertia from the falling velocity and its associated acceleration. Due to that, the drag force must be equal to the submerged weight of the particle, and any lift forces are caused by turbulence.

When assessing a two-phase flow, it is necessary to assess separate timeframes, i.e., the particle (t_p) and integral turbulence timeframes (t_t). As shown by Equation (8), F_D can be separated into two factors for laminar and turbulent flows [20]:

$$F_D = \rho_s \bar{c}\left(\frac{\omega_0}{t_p} + \frac{\Delta w_d}{t_t}\right) \tag{8}$$

where $\bar{c}$ is the mean concentration, ω_0 is the fall velocity of sediment, and Δw_d is the velocity flux difference between the two phases and is considered proportional to ω_0. Particle crowding can have a significant impact on the particle timeframe. Research has shown that particle flocculation is a function of Reynolds number; however, it has also been defined as a constant in some studies [14]. Nonetheless, for dilute flows, the gravitational particle timeframe can be defined as [20]:

$$t_p = \frac{\omega_0}{g\left(1 - \frac{\rho_f}{\rho_s}\right)} \tag{9}$$

where ρ_s and ρ_f are the sediment and fluid densities, respectively. (t_p) is often associated with Stokes' drag. Stokes' number, St_b, describes the relationship between the relative particle velocity and the particulate timeframe. When a particle moves within turbulent flow, the fluid velocity fluctuation and formation of vortices exert a force on the particle and causes longer t_t, where this timeframe can be determined by [13,14,20,31]:

$$t_t = \frac{t_p}{St_b - 1} \tag{10}$$

For light sediment, i.e., sediment moves in equilibrium with the fluid, the flow can effectively be treated as a uniform medium. In this case, Stokes' number will be very small. When a particle experiences considerable inertial forces, St_b will be close to one. Therefore, substituting Equations (9) and (10) into Equation (8) yields:

$$F_D = \bar{c}\rho_s g\left(1 - \frac{\rho_f}{\rho_s}\right) + \bar{c}\rho_s(St_b - 1)\frac{\Delta w_d}{t_p} \tag{11}$$

The first term on the right-hand side of Equation (11) refers to the laminar drag force, and the second term refers to the turbulent drag force.

2.3. *Lift Force*

Conventionally lift is considered to act upwards from the horizontal plane opposing gravity. However, it is practical to consider lift acting perpendicular to the streamwise direction [32,33]. Besides lift, the cohesive force between particles can also cause the drag between them to create a change in suspended sediment transport. However, the cohesive force is more effective within bedload and not suspended load. This is due to the fact that the distance between particles in bedload is much shorter than that in suspended load, which is usually very scattered. Due to this, the cohesive force between suspended particles is not a determinant factor of its behavior as compared to bedload [34].

Commonly, lift force is considered to be composed of two parts: hydrodynamic lift and turbulent diffusion. One of the limits of original Rouse suspended sediment model is that it suggests an infinite concentration at the bed and zero concentration at the surface. It has been shown that a lower concentration corresponds to a higher horizontal velocity, where this is represented in Figure 3. The importance of this factor is not fully understood; however, it is known that hydrodynamic lift is the result of the concentration variation and thus velocity differential across the sediment particle from the lower to upper suspension region (Figure 3 presents the pattern of velocity differential across a particle in non-laminar flow). Additionally, the displacement of fluid around the body causes a pressure distribution across the particle, which consequently causes a velocity gradient. Therefore, a lift force is acted on the particle to equalize the pressure gradient [35].

Figure 3. Uplift due to shear stresses on a particles surface. Reprinted with permission from [36]. 2021 Hanmaiahgari, P.R.

Turbulence diffusion is the mixing due to the turbulent velocity fluctuation and vortices. Many studies have been conducted to understand the causes and effects of turbulence. It has been assessed statistically as an overlaying function of the time-averaged velocity [20–22]. The wake of a sediment particle (shown in Figure 3) is often considered to cause turbulent mixing due to the velocity fluctuations. Generally, this turbulent mixing is associated with large vortices as small eddies relative to the sediment size do not contain sufficient energy to lift the sediment particle. By this consideration, turbulence can be correlated with shear velocity [23]. Additionally, turbulent diffusion primarily occurs away from near-bed flow region. This is due to the lower horizontal velocities apparent at the region, as sediment concentration is generally higher.

Within the research of sediment-laden flow, Zhong et al. [37] concluded that near the bed, there is no sufficient space for vortices to form within the fluid. Additionally, it is also found that a lower eddy viscosity is correlated with a lower bed slope, and thus, density stratification is more prevalent [21,33,38]. This supports the notion that higher velocities produce more turbulence. Furthermore, Figure 4 shows that the strongest eddy viscosity occur within the center of the flow column whereas the highest velocity occurs at the surface of the flow [39].

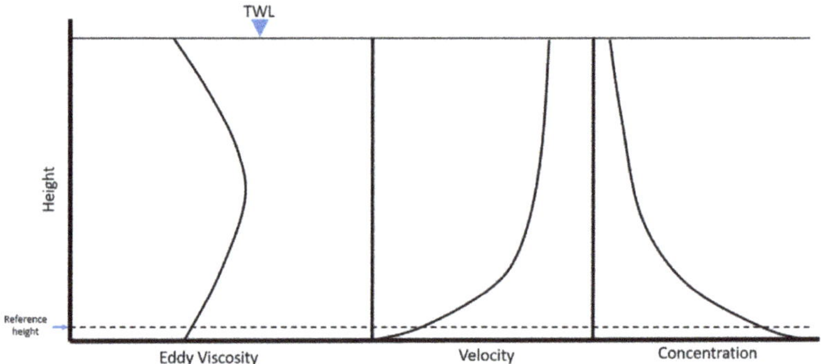

Figure 4. Eddy viscosity, velocity, and concentration distribution profile (Adapted from Liu and Nayamatullah [39]).

Upward turbulent velocities are countered by gravity and downward turbulent bursting mechanism, i.e., sweep. Huang et al. [9] suggest that settling velocity is composed of these two components in the form of

$$\Delta w_d = \Delta w_g + \Delta w_t \tag{12}$$

where Δw_d is the sediment fluctuation, Δw_g is the settling flux due to gravity, and Δw_t is the turbulent diffusive flux. The settling flux due to gravity is well understood, where it is influenced by the downward gravitional acceleration. However, as stated previously, the diffusive flux is more chaotic and random in nature and is thus more difficult to model. Previous studies have described this by analyzing a vertical unit sectional area and describing the turbulent velocity dependent on the concentration differential at the top and bottom of the sectional area of flow channel [9,24]. In an alternative and simpler way of interpreting this, the relative velocity of a particle is described by the difference between the fluid and particle velocity take in account any drift velocity [13,14,25]:

$$u_r = u_f - u_s \pm u_d \tag{13}$$

where $u_r = u_f - u_s \pm u_d$ is the relative velocity, $u_r = u_f - u_s \pm u_d$ is the fluid velocity, $u_r = u_f - u_s \pm u_d$ is the solid velocity, and $u_r = u_f - u_s \pm u_d$ is the drift velocity. The difficulty then lies in determining u_d. As stated previously, it is known that drift velocity is a result of vortices and fluid turbulence. The drift velocity can be further defined as [14]:

$$u_d = -D\left(\frac{1}{c}\frac{dc}{dy} - \frac{1}{1-c}\frac{d(1-c)}{dy}\right) \quad (14)$$

where D is the diffusion tensor and is considered to be dependent on the Reynolds stress, which, in turn, is related to the vortex viscosity [25]:

$$D = -t_t\overline{v'_f v'_s} = -\frac{t_p}{St_b - 1}\overline{v'_f v'_s} \quad (15)$$

where $\overline{v'_f v'_s}$ is the Reynolds stress and it can be calculated by:

$$-\overline{v'_f v'_s} = \mu\left(\frac{\partial u_f}{\partial y} + \frac{\partial u_s}{\partial y}\right) - \frac{2}{3}E_k\delta \quad (16)$$

where μ is the eddy viscosity, E_k is the kinetic energy, and δ is the Kronecker delta.

Furthermore, research inspecting the velocity lag of a particle within flow arising from the drag force shows that the relationship for the lag velocity in dilute flow is defined as [40]:

$$\frac{u_l}{u_*} = \left[\sqrt{\left(\frac{8v}{u_* d}\right)^{2/1.5} + \left(\frac{2(h-y)}{h}\right)^{1/1.5}} - \left(\frac{16v}{u_* d}\right)^{1/1.5}\right]^{1.5} \quad (17)$$

where u_* is the shear velocity, and v is the kinematic viscosity. It is worth noting that Equation (17) is only valid when assessing dilute flow. For alternative cases, i.e., concentrated flow, the constants and exponents will be varying [18]. As u_l is considered the difference between u_f and u_s, it can be assumed that:

$$\frac{u_r}{u_*} = \frac{u_l}{u_*} + \frac{u_d}{u_*} \quad (18)$$

Therefore, inputting Equations (10), (14), (15) and (17) into Equation (18) gives a conclusive definition of the dimensionless relative velocity:

$$\frac{u_r}{u_*} = \frac{u_l}{u_*} + \frac{\overline{v'_f v'_s}\frac{\omega_0}{u_*}}{gc(1-c)(St_b - 1)\left(1 - \frac{\rho_f}{\rho_s}\right)}\frac{dc}{dy} \quad (19)$$

For a single particle, its effective lift force can be defined as [37,41]:

$$F_L = C_L\frac{4\pi}{3}\rho_f d^3 u_r \frac{du_f}{dy} \quad (20)$$

Alternatively, the lift force per unit of mass is described by [8]:

$$L_F = \frac{6F_L}{\rho_s \pi d^3} \quad (21)$$

Thus, the total lift force exerted on the solid phase can be calculated by:

$$M_{sy} = c\rho_s L_F \quad (22)$$

Inputting Equation (19) into Equation (20) determines the lift force acting upon a single particle due to the turbulent diffusion as:

$$F_L = C_L \frac{4\pi}{3} \rho_f d^3 \left(u_l + \frac{\overline{v'_f v'_s} \omega_0}{gc(1-c)(St_b - 1)\left(1 - \frac{\rho_f}{\rho_s}\right)} \frac{dc}{dy} \right) \frac{du_f}{dy} \tag{23}$$

Theodore von Karman determined that hydraulic flow within close proximity to the wall follow a velocity distribution proportional to the logarithm of the distance from the wall [42]:

$$\frac{u}{u_*} = \frac{1}{\kappa} ln\left(\frac{y}{h}\right) + A \tag{24}$$

where κ is the von Karman constant, and A is the log-law constant. However, this is generally only for boundaries close to the bed, and it loses its accuracy within the upper flow region, especially when suspended sediment is detected. The effect of a boundary causes shear on the particles, which results in secondary currents within the flow. Considerable research has been undertaken to study the effect of secondary vortices and wake within the near bed region [16,21]. In Equation (24), when the law of wake has been considered, the constant A can be defined as [16]:

$$A = \frac{2\Pi}{\kappa} \left(3\left(\frac{y}{h}\right)^2 - 2\left(\frac{y}{h}\right)^3 \right) \tag{25}$$

where Π is Cole's wake parameter. Therefore, inputting Equation (23) into Equation (22) produces:

$$\frac{u}{u_*} = \frac{1}{\kappa} ln\left(\frac{y}{h}\right) + \frac{2\Pi}{\kappa} \left(3\left(\frac{y}{h}\right)^2 - 2\left(\frac{y}{h}\right)^3 \right) \tag{26}$$

Substituting Equation (23) into Equation (21) and then inserting into Equation (22) with the inclusion of Equation (26) defines the total lift force, which occurs on the sediment phase from the turbulent velocity and secondary currents:

$$M_{sy} = c\gamma_2 \rho_s \frac{u_*^2}{h} \left[\frac{h}{y} + 12\Pi\left(\frac{y}{h}\right)\left(1 - \frac{y}{h}\right) \right] \left\{ \frac{u_l}{u_*} + \frac{\varnothing}{c(1-c)} \frac{\partial c}{\partial y} \right\} \tag{27}$$

where γ_2 and $\varnothing$ are constants defined as

$$\gamma_2 = \frac{8 C_L \rho_f}{k \rho_s} \tag{28}$$

and

$$\varnothing = \frac{\frac{\omega_0}{u_*}}{g(St_b - 1)\left(1 - \frac{\rho_f}{\rho_s}\right)} \mu \tag{29}$$

2.4. Turbulent Bursting

When mathematically modelled, a time-averaged mean velocity is often employed to represent the suspended solid phase. However, more recently, studies have shown that instantaneous turbulent velocities at the near bed region can have significant impact on the overall sediment concentration distribution in flow [22,43]. These instantaneous flow properties are also responsible for solid particles to be entrained into the suspended region from the bed-load, which can be analyzed as turbulent bursting. Additionally, the sediment particles at the bed experience higher velocity gradients, thus generating larger lift acts on the particle. In some instances, this can cause scouring of the sea-bed, as well as unexpected deposition, resulting in risk to marine infrastructure as well as aquatic life [44–46].

Turbulence is hard to consider in modelling terms. Due to the nature of fluid, a vortex will impose a force on the surrounding fluid producing smaller vortices until the vortex's kinetic energy is insufficient to overcome the viscosity of the fluid. Turbulence can be broken down in to three main stages: production, transfer, and dissipation, otherwise known as integral scale, inertial scale, and Kolmogorov scale, respectively, as shown in Figure 5.

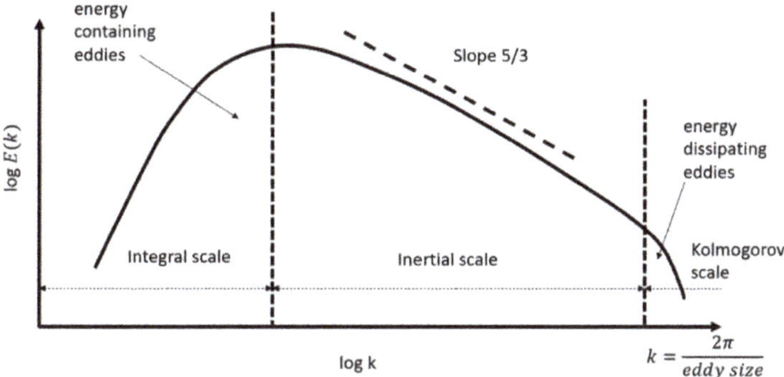

Figure 5. Turbulence eddy scaling. Reprinted with permission from [41]. 2021 Hanmaiahgari, P.R.

A vortex can be defined by its diameter, velocity, and timeframe. Andrey Kolmogorov discovered that within the transfer region (i.e., the inertial scale), the kinetic energy is proportional to the diameter of the eddy, where it can be described as [47]

$$K \cdot E_v = C_k \varepsilon^{\frac{2}{3}} k^{\frac{-5}{3}} \tag{30}$$

where C_k is the Kolmogorov constant, ε is the dissipation rate, and k is the wave number in a function of the eddy size. The eddy scale is defined by varying characteristics dependent upon the stage of turbulence, i.e., the integral scale is a function of kinetic energy and dissipation, the inertial scale is a function of dissipation and length, and the Kolmogorov scale is a function of dissipation and viscosity.

It is understood that a critical velocity value criterion is required to meet for sediment suspension to occur, and sediment entrained within the fluid phase below the critical velocity is considered as wash-load [24,48,49]. Salim et al. [22] studied the sediment suspension at the near-bed level in the Australian ocean for locations with current and wave velocities lower than the critical value. They found that despite the relatively low recorded flow velocity, sediment suspension still took place due to turbulent bursting. Similarly, Tsai and Huang [43] investigated the advection-diffusion method of modelling sediment concentration and concluded the instantaneous velocity fluctuations are mainly due to turbulent ejection.

Figure 6 shows a comparison of field data at the seabed for study by Salim et al. [22]. From Figure 6, it is possible to see that a higher mean velocity value commonly corresponds to higher backscatter and TKE shear stress. Timeframes (i), (ii), and (iii) in the figure all represent sediment suspension occurrences. This was considered sensible, as the velocity was at or above the critical value for their study where sediment suspension occurred. However, conversely at timeframes TS-1 and TS-2, it was observed that there was a high value for the TKE shear stress and backscatter, whereas there was a low value for the streamwise velocity, i.e., below the critical velocity value. This increase in the TKE shear stress suggests the formation of vortices, which could lift the sediment into the suspension zone, and it is supported by the high backscatter reading, proving sediment suspension due to bursting's sweep events.

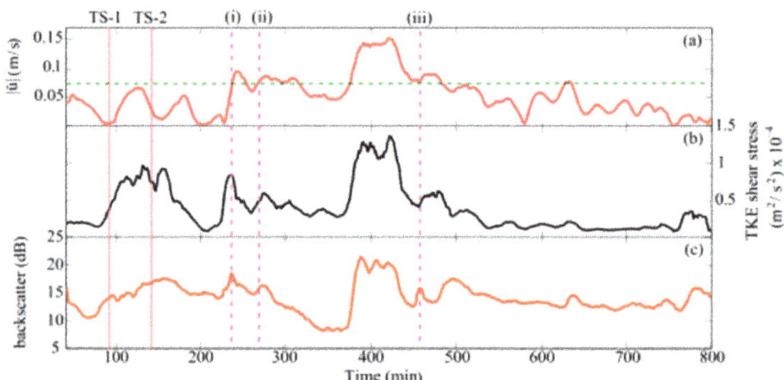

Figure 6. Time series showing: (**a**) 1 min averaged velocity streamwise velocity, (**b**) turbulent kinetic energy (TKE) shear stress, and (**c**) backscatter. The horizontal line shown in (**a**) represents the critical velocity value for suspension. Reprinted with permission from [22]. 2021 Hanmaiahgari, P.R.

In recent experimental advances, particle-imaging velocimetry (PIV) has been used for suspended particle investigation in various wave-induced and turbulent flow conditions [50,51]. It has been evidenced that PIV is a good method to capture the suspended sediment dynamics as long as relatively good resolution of camera has been deployed. By this finding, one can project that the suspended sediment transport behavior can be accurately captured in real-world events in the future.

2.5. Continuity Equations and Modelling Studies

Within common solid-liquid flow modelling, the mass conservation system is usually assumed, while its acceleration can be variable. The coordinate system of a single particle is represented in Figure 7.

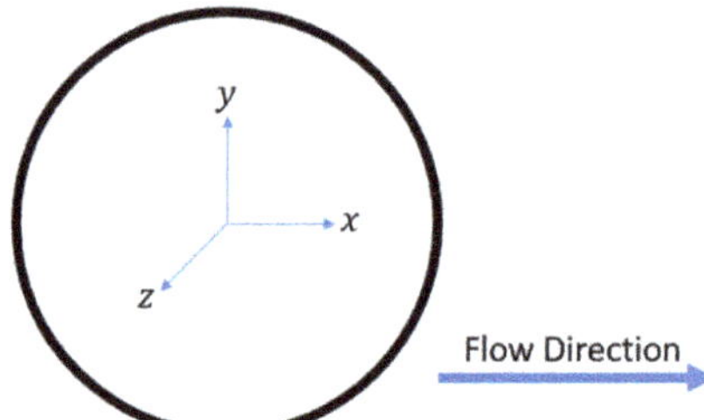

Figure 7. Polar coordinate system.

Within a fluid flow convection, acceleration can be considered by continuity rule over time, as defined by [52]:

$$a = u\frac{du}{dx} + \frac{du}{dt} \tag{31}$$

Equation (31) can be decomposed into constituent directions, which include each spatial plane. Thus, the vertical flow acceleration can be expressed by expanding Equation (31) to give:

$$a_y = g_y = \frac{dv}{dt} + v\frac{du}{dx} + v\frac{dv}{dy} + v\frac{dw}{dz} \tag{32}$$

However, within a two-phase turbulent flow, there are other resulting forces acting on the particle, which cause velocity fluctuations. Stokes and Navier individually derived the Navier-Stokes equations stemming from Equation (17) to incorporate the additional flow effects within viscous flow regimes.

In general, suspended load transport is derived from the continuity equations and a balancing of forces acting on the solid phase [8,15,20,53–55]. A closed channel model assumes no particle enters or exits the flow. In a mathematical model, the respective interactions and boundaries are then considered to give an overall expression for the sediment profile.

The mass conservation equation is derived from the Navier-Stokes equations in accordance with the closed system assumption and can be expressed as:

$$c\left(\frac{d}{dt}+\frac{du_s}{dx}+\frac{dv_s}{dy}+\frac{dw_s}{dz}\right)=0 \tag{33}$$

where c is the mean concentration of the solid phase. Similarly, the conservation of momentum equation is given as:

$$c\left(\frac{\partial v_s}{\partial t}+v_s\frac{\partial u_s}{\partial x}+v_s\frac{\partial v_s}{\partial y}+v_s\frac{dw_s}{dz}\right)=cg_y-\frac{c}{\rho_s}\frac{\partial p_f}{\partial y}+\frac{1}{\rho_s}\frac{\partial \sigma_s}{\partial y}+\frac{1}{\rho_s}F_{sy} \tag{34}$$

where p_f is the fluid flow pressure, σ_s is the tensor stress arising from particle interactions, and F_{sy} is the vertical forces arising due to phase interactions. Considering a 2D flow, where the flow across channel width is considered having ignorable change, the dz term can be removed from Equation (34). Furthermore, within steady-uniform flow assumption, one can obtain:

$$\frac{d}{dt}=\frac{d}{dx}=0 \tag{35}$$

$$g_y=g \tag{36}$$

Therefore, the only varying values are given in the streamwise-vertical plane. Additionally, for dilute flow conditions, the particle-particle interactions are negligible:

$$\sigma_s=0 \tag{37}$$

Thus, if Equations (35)–(37) are inserted into Equation (34), the vertical momentum for the solid phase can be defined as:

$$c\left(v_s\frac{\partial v_s}{\partial y}\right)=cg-\frac{c}{\rho_s}\frac{\partial p_f}{\partial y}+\frac{1}{\rho_s}F_{sy} \tag{38}$$

Within turbulent flow, Reynolds decomposition is a mathematical method used to isolate the velocity fluctuation values from the averaged value (as presented by Figure 8), given by [56]:

$$u=\overline{u}+u\prime \tag{39}$$

where $\overline{u}$ is the time-averaged velocity, and u' is the velocity flux.

Applying Reynolds decomposition to the solid phase by inputting Equation (39) into Equation (38) yields:

$$c\left(\overline{v}+v_s'\right)\left(\overline{v}+v_s'\right)\frac{\partial}{\partial y}=cg-\frac{c}{\rho_s}\frac{\partial p_f}{\partial y}+\frac{1}{\rho_s}F_{sy} \tag{40}$$

Expanding and simplifying Equation (40) gives:

$$c\left(\overline{v}_s+2v_s'\right)\frac{\partial \overline{v}}{\partial y}+cv_s'^2\frac{\partial}{\partial y}=cg-\frac{c}{\rho_s}\frac{\partial p_f}{\partial y}+\frac{1}{\rho_s}F_{sy} \tag{41}$$

Additionally, applying the assumptions of Equations (35) and (40) to Equation (33) gives:

$$c\left(\overline{v_s}+v_s'\right)\frac{\partial}{\partial y}=0 \tag{42}$$

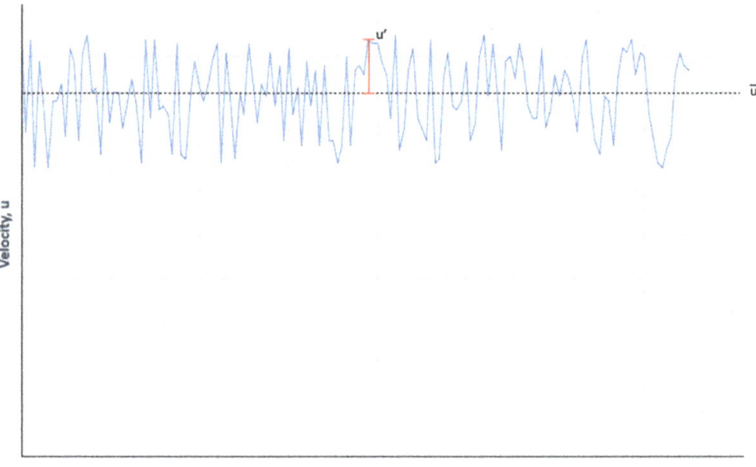

Figure 8. Depiction of turbulent velocity fluctuation.

Equation (42) can be solved through integration to obtain:

$$c\left(\overline{v_s} + v_s'\right) = 0 \tag{43}$$

The first term on the left-hand side of Equation (40) is usually considered negligible, and thus, rewriting and multiplying by ρ_s, Equation (40) becomes:

$$c\frac{\partial p_f}{\partial y} - c\rho_s g - \rho_s c\frac{\partial v_s'^2}{\partial y} + F_{sy} = 0 \tag{44}$$

The first term accounts for the pressure variation due to increasing depth.

The hydrostatic pressure is a result of the weight of the fluid and can be defined as:

$$\frac{\partial p_f}{\partial y} = -\rho_f g \tag{45}$$

Two main interactions that have been discussed within this report are drag and lift. These forces can be combined by adding Equations (11) and (29) together to produce:

$$F_{sy} = c\rho_s g\left(1 - \frac{\rho_f}{\rho_s}\right) + c\rho_s(St_b - 1)\frac{\Delta w}{t_p} + c\gamma_2\rho_s\frac{u_*^2}{h}\left[\frac{h}{y} + 12\Pi\left(\frac{y}{h}\right)\left(1 - \frac{y}{h}\right)\right]\left\{\frac{u_l}{u_*} + \frac{\varnothing}{c(1-c)}\frac{\partial c}{\partial y}\right\} \tag{46}$$

Therefore, inputting Equations (43) and (44) into Equation (42) produces:

$$\begin{aligned} c\rho_f g - c\rho_s g - \rho_s c\tfrac{\partial v_s'^2}{\partial y} + c\rho_s g\left(1 - \tfrac{\rho_f}{\rho_s}\right) + c\rho_s(St_b - 1)\tfrac{\Delta w}{t_p} \\ + c\gamma_2\rho_s\tfrac{u_*^2}{h}\left[\tfrac{h}{y} + 12\Pi\left(\tfrac{y}{h}\right)\left(1 - \tfrac{y}{h}\right)\right]\left\{\tfrac{u_l}{u_*} + \tfrac{\varnothing}{c(1-c)}\tfrac{\partial c}{\partial y}\right\} = 0 \end{aligned} \tag{47}$$

Rearranging for sediment concentration finally gives:

$$v_s'^2\frac{\partial c}{\partial y} - c\gamma_2\frac{u_*^2}{h}\frac{\varnothing}{c(1-c)}\frac{\partial c}{\partial y} = cg\frac{\rho_f}{\rho_s}\left(1 - \frac{\rho_f}{\rho_s}\right) + c(St_b - 1)\frac{\Delta w}{t_p} + c\gamma_2 x\frac{u_*^2}{h}\left[\frac{h}{y} + 12\Pi\left(\frac{y}{h}\right)\left(1 - \frac{y}{h}\right)\right]\left\{\frac{u_l}{u_*}\right\} \tag{48}$$

Equation (48) is a comprehensive and inclusive approach to mathematically represent the sediment concentration distribution profile. It accounts for the velocity lag due to particle inertia associated with the drag force as well as the velocity fluctuations arising due to the turbulent flow and formation of vortices. Recently, the non-local theory through the fractional advection–diffusion equation (fADE) has also been proven to represent suspension sediment concentration well [57]. In terms of numerical modelling, various lattice Boltzmann (LB) [58,59] and smoothed particle hydrodynamics (SPH) models [60,61] have also been investigated for their representation toward flow with suspended and bed loads, and great success has been achieved in terms of accuracy in reproducing the experimental and field data.

The above-mentioned final equation (Equation (48)) can also be simplified. If small particles are measured, then it is justifiable to assume that the solid phase moves in equilibrium with main flow, and there is no velocity lag. Therefore, St_b can be taken as zero. Additionally, for fine smooth particles, there would be very little shear stress on the surface of the particle to give rise to lift forces. Finally, if the formation of eddies due to turbulence is ignored, these assumptions in Equation (48) can give rise to the following:

$$D_{yy}\frac{dc}{dy} = -ct_p\left(1 - \frac{\rho_f}{\rho_s}\right)g \tag{49}$$

Inputting Equation (9) into Equation (49) gives:

$$D_{yy}\frac{dc}{dy} = -\omega_0 c \tag{50}$$

This equates to Fick's Law, which defines diffusion theory and states that diffusion from an area of high concentration to an area of low concentration should be balanced by settling velocity term.

3. Conclusions

A wide range of research has been reviewed to provide a holistic understanding of sediment concentration profiles, suspension, and transport for practical use when modelling hydrodynamic flows. Drag on the sediment particle has been assessed using the Reynolds number approach. It has been highlighted that the turbulent flow condition plays a significant role on the drag force experienced by the particle. This is dependent upon the shear forces between the solid and fluid phase. A study of velocity lag has also been considered by utilizing the Stokes number, which is a relationship between the particle velocity and particle transport timeframe. Research has supported the notion that drag force is significantly affected by the turbulent flow condition.

The idea of velocity fluctuations and vortex mixing has been studied. Velocity fluctuations in the upward direction give lift to the sediment particle. A general approach of the time-averaged velocity fluctuations has been considered. The effects of turbulent bursting in the upward motion close to the bed has also been reviewed. Within low flow conditions, where there are no sufficient velocities for suspension action, a time-averaged velocity method is not sufficient to describe particle suspension. In this case, research has found that upward eddies can alternatively lift particles into the flow column. The continuity equations have been investigated to assess the variance of acceleration with positional displacement over time. This has been expanded by researchers to incorporate other forces that act upon the sediment particle such as the fluid-induced lift and drag forces, forces arising from turbulence, particle inertia, and pressure gradient.

Conclusively, as found from this review study, there are a wide range of flow parameters that affect the suspended sediment transport. Due to this, reasonable assumptions and simplifications have been proposed to ease the mathematical modelling burden. In-depth suspended sediment transport study can benefit flow modelling under marine and water infrastructure influences. For bridge-pier or abutment-induced flow, the flow turbulence will be significant. This will impact the characteristic of scour and suspended sediment transport. Thus, in future studies, a more inclusive research strategy will be needed to accurately model the suspended solid behavior within flow to foresee any environmental problem that could be caused in man-made or natural channels.

Author Contributions: J.T.W.: writing—original draft preparation, review and editing, funding; J.H.P.: writing—original draft preparation, review and editing; S.K.: writing—review and editing; P.R.H.: writing—review and editing; M.P.: writing—review and editing; A.S.: writing—review and editing; M.A.K.: writing—review and editing; and A.W.: writing—review and editing. All authors have read and agreed to the published version of the manuscript.

Funding: This research received no external funding.

Data Availability Statement: Not applicable.

Acknowledgments: The author, Jaan H. Pu, acknowledges a travel grant from the Faculty of Engineering and Informatics, University of Bradford, to explore potential research collaboration opportunity.

Conflicts of Interest: The authors declare no conflict of interest.

References

1. Kundu, S.; Ghoshal, K. Effects of Secondary Current and Stratification on Suspension Concentration in an Open Channel Flow. *Environ. Fluid Mech.* **2014**, *14*, 1357–1380. [CrossRef]
2. Pu, J.H. Turbulent rectangular compound open channel flow study using multi-zonal approach. *Environ. Fluid Mech.* **2018**, *19*, 785–800. [CrossRef]
3. Pu, J.H.; Pandey, M.; Hanmaiahgari, P.R. Analytical modelling of sidewall turbulence effect on streamwise velocity profile using 2D approach: A comparison of rectangular and trapezoidal open channel flows. *J. Hydro-Environ. Res.* **2020**, *32*, 17–25. [CrossRef]
4. Pu, J.H. Velocity profile and turbulence structure measurement corrections for sediment transport-induced water-worked bed. *Fluids* **2021**, *6*, 86. [CrossRef]
5. Pu, J.H.; Wei, J.; Huang, Y. Velocity distribution and 3D turbulence characteristic analysis for flow over water-worked rough bed. *Water* **2017**, *9*, 668. [CrossRef]
6. Hanmaiahgari, P.R.; Gompa, N.R.; Pal, D.; Pu, J.H. Numerical modeling of the sakuma dam reservoir sedimentation. *Nat. Hazards* **2018**, *91*, 1075–1096. [CrossRef]
7. Pu, J.H.; Huang, Y.; Shao, S.; Hussain, K. Three-gorges dam fine sediment pollutant transport: Turbulence SPH model simulation of multi-fluid flows. *J. Appl. Fluid Mech.* **2016**, *9*, 1–10. [CrossRef]
8. Zhong, D.; Wang, G.; Sun, Q. Transport Equation for Suspended Sediment Based on Two Fluid Model of Solid/Liquid Two-Phase Flows. *J. Hydraul. Eng.* **2011**, *137*, 530–542. [CrossRef]
9. Huang, S.H.; Sun, Z.L.; Xu, D.; Xia, S.S. Vertical Distribution of Sediment Concentration. *J. Zhejiang Univ. Sci.* **2003**, *9*, 1560–1566. [CrossRef]
10. Hsu, T.J.; Jenkins, J.T.; Liu, P.L.F. On Two-Phase Sediment Transport: Dilute Flow. *J. Geophys. Res.* **2003**, *108*, 2–14. [CrossRef]
11. Rouse, H. *Modern Conceptions of the Mechanics of Fluid Turbulence*; American Society of Civil Engineers: Reston, VI, USA, 1936; Volume 62, pp. 21–64.
12. Fick, A. On Liquid Diffusion. *J. Membr. Sci.* **1995**, *100*, 33–38. [CrossRef]
13. Kundu, S.; Ghoshal, K. A mathematical model for type II profile of concentration distribution in turbulent flows. *Environ. Fluid Mech.* **2017**, *17*, 449–472. [CrossRef]
14. Greimann, B.P.; Holly, F.M., Jr. Two-phase flow analysis of concentration profiles. *Hydraul. Eng.* **2001**, *127*, 753–762. [CrossRef]
15. Jha, S.K.; Bombardelli, F.A. Two-Phase Modelling of Turbulence in Dilute Sediment Laden Open-Channel Flow. *Environ. Fluid Mech.* **2009**, *9*, 237. [CrossRef]
16. Kundu, S.; Ghoshal, K. Explicit formulation for suspended concentration distribution with near-bed particle deficiency. *Powder Technol.* **2013**, *253*, 429–437. [CrossRef]
17. Jha, S.K.; Bombardelli, F.A. Toward two-phase flow modelling of nondilute sediment transport in open channels. *J. Geophys. Res.* **2010**, *115*, F3.
18. Pu, J.H.; Wallwork, J.T.; Khan, M.A.; Pandey, M.; Pourhahbaz, H.; Satyanaga, A.; Hanmaiahgari, P.R.; Gough, T. Flood suspended sediment transport: Combined modelling from dilute to hyper-concentrated flow. *Water* **2021**, *13*, 379.
19. Goree, J.C.; Keetels, G.H.; Munts, E.A.; Bugdayci, H.H.; Rhee, C.V. Concentration and velocit profiles of sediment-water mixtures using the drift flux model. *Can. J. Chem. Eng.* **2016**, *94*, 1048–1058.
20. Toorman, E.A. Vertical Mixing in the Fully Developed Layer of Sediment-Laden Open-Channel Flow. *J. Hydraul. Eng.* **2008**, *134*, 1225–1235. [CrossRef]
21. Moodie, A.J.; Nittrouer, J.A.; Ma, H.; Carlson, B.N.; Wang, Y.; Lamb, M.P.; Parker, G. Suspended-sediment induced stratification inferred from concentration and velocity profile measurements in the Yellow River, China. *Water Resour. Res.* **2019**, e2020WR027192. [CrossRef]
22. Salim, S.; Pattiaratchi, C.; Tinoco, R.O.; Jayaratne, R. Sediment resuspension due to near-bed turbulent effects: A deep sea case study on the northwest continental slope of Western Australia. *J. Geophys. Res. Ocean.* **2018**, *123*, 7102–7119. [CrossRef]

23. Einstein, H.A. *The Bed-Load Function for Sediment Transport in Open Channel Flows*; Departmetn of Agriculture: Washington, DC, USA, 1950.
24. Drew, D.A. Mathematical modelling of two-phase flow. *Fluid Mech.* **1983**, *15*, 261–291. [CrossRef]
25. Ni, J.R.; Wang, G.Q.; Borthwick, A.G.L. Kinetic Theory for Particles in Dilute and Dense Solid-Liquid Flows. *J. Hydraul. Eng.* **2000**, *126*, 893–903. [CrossRef]
26. Sun, Z.; Zheng, H.; Xu, D.; Hu, C.; Zhang, C. Verticle Concentration Profile of Nonuniform Sediment. *Int. J. Sediment Res.* **2021**, *36*, 120–126. [CrossRef]
27. Soo, S.L. *Fluid Dynamics of Multiphase Systems*; Blaisdell Publishing Company: Waltham, MA, USA, 1967.
28. Clift, R.; Grace, J.R.; Weber, M.E. *Bubbles Drops and Particles*; Academic Press: New York, NY, USA, 1978.
29. Van Nierop, E.A.; Luther, S.; Bluemink, J.J.; Magnaudet, J.; Prosperetti, A.; Lohse, D. Drag and Lift Force on Bubbles in a Rotating Flow. *J. Fluid Mech.* **2007**, *571*, 439–454. [CrossRef]
30. Cheng, N.S. Comparison of formulas for drag coefficient and settling velocity of spherical particles. *Powder Technol.* **2009**, *189*, 395–398. [CrossRef]
31. Pal, D.; Ghoshal, K. Hydrodynamic Interaction in Suspended Sediment Distribution of Open Chanel Turbulent Flow. *Appl. Math. Model.* **2017**, *49*, 630–646. [CrossRef]
32. Liu, D.Y. *Fluid Dynamics of Two-Phase Systems*; Chinese Higher Education Press: Beijing, China, 1993.
33. Hamil, L. *Understanding Hydraulics*; Palgrave MacMillan: London, UK, 2001.
34. Jain, R.K.; Kothyari, U.C. Cohesion influences on erosion and bed load transport. *Water Resour. Res.* **2009**, *45*, 6. [CrossRef]
35. Chanson, H. *The Hydraulics of Open Channel Flow*; Hodder Headline Group: London, UK, 1999.
36. Kurose, R.; Komori, S. Drag and Lift Force on a Rotating Sphere in Linear Shear Flow. *J. Fluid Mech.* **1999**, *384*, 183–206. [CrossRef]
37. Zhong, D.; Zhang, L.; Wu, B.; Wang, Y. Velocity Profile of Turbulent Sediment-Laden Flows in Open-Channels. *Int. J. Sediment Res.* **2015**, *30*, 285–296. [CrossRef]
38. Wright, S.; Parker, G. Density Stratification Effects in Sand-Bed Rivers. *J. Hydraul. Eng.* **2004**, *130*, 783–795. [CrossRef]
39. Liu, X.; Nayamatullah, M. Semianalytical solutions for one-dimensional unsteady nonequilibrium suspended transport in channels with arbitrary eddy viscosity distribution and realistic boundary conditions. *J. Hydraul. Enigneering* **2014**, *5*, 1–10. [CrossRef]
40. Cheng, N.S. Analysis of velocity lag in sediment-laden open channel flows. *J. Hydraul. Eng.* **2004**, *130*, 657–666. [CrossRef]
41. Chen, X.; Li, Y.; Niu, X.; Li, M.; Chen, D.; Yu, X. A general two-phase turbulent flow model applied to the study of sediment transport in open channels. *Int. J. Multiph. Flow* **2011**, *37*, 1099–1108. [CrossRef]
42. Bradshaw, P.; Huang, G.P. The law of the wall in turbulent flow. *Math. Phys. Sci.* **1995**, *451*, 165–188.
43. Tsai, C.W.; Huang, S.H. Modeling suspended sediment transport under influence of turbulence ejection and sweep events. *Water Resour. Res.* **2019**, *55*, 5379–5393. [CrossRef]
44. Leckie, S.H.F.; Mohr, H.; Draper, S.; McLean, D.L.; White, D.J.; Cheng, L. Sedimentation-induced Burial os Subsea Pipelines: Observation from Field Data and Laboratory Experiments. *Coast. Eng.* **2016**, *114*, 137–158. [CrossRef]
45. Isadinia, E.; Heidarpour, M.; Schleiss, A.J. Investigation of turbulence flow and sediment entrainment around bridge peir. *Stoch. Environ. Res. Risk Assess.* **2013**, *27*, 1303–1314. [CrossRef]
46. Sinha, N. *Towards RANS Parameterized of Vertical Mixing by Langmuir Turbulence in Shallow Coastal Shelves*; University of South Florida Scholar Commons: Sarasota, FL, USA, 2013.
47. Kolmogorov, A.N. The Local Structure of Turbulence in Incompressible Viscious Fluids at Very Large Reynolds Numbers. *Math. Phys. Sci.* **1991**, *434*, 9–13.
48. Baumert, H.Z.; Simpson, J.; Sundermann, J. *Marine Turbulence: Theories, Observations and Models*; Cambridge University Press: Cambridge, UK, 2005.
49. Wang, Y.Y.; He, Q.; Liu, H. Variations of Near-Bed Suspended Sediment Concentration in South Passage of the Changjiang Estuary. *J. Sediment Res.* **2009**, *6*, 6–13.
50. Stachurska, B.; Staroszczyk, R. Loboratory study of suspended sediment dynamics Over a mildly sloping sandy bed. *Oceanologia* **2019**, *61*, 350–367. [CrossRef]
51. Li, G.; Gao, Z.; Li, Z.; Wang, J.; Derksen, J.J. Particle-Resolved PIV Experiments of Solid-Liquid Mixing in a Turbulent Stirred Tank. *Am. Inst. Chem. Eng. J.* **2017**, *64*, 389–402. [CrossRef]
52. Batchelor, G. *An Introduction to Fluid Mechanics*; Cambridge University Press: Cambridge, UK, 1967.
53. Gavrilov, A.A.; Finnikov, K.A.; Ignatenko, Y.S.; Bocharov, O.B.; May, R. Drag and lift forces acting on a sphere in shear flow of power-law fluid. *J. Eng. Thermophys.* **2018**, *27*, 474–488. [CrossRef]
54. Ali, S.Z.; Dey, S. *Mechanics of Advection of Suspended Particles in Turbulent Flow*; The Royal Society Publishing: Kharagpur, India, 2016.
55. Cui, H.; Singh, V.P. Suspended sediment concentration in open channel using tsallis entropy. *J. Hydraul. Eng.* **2014**, *19*, 966–977. [CrossRef]
56. Bose, S.K.; Dey, S. Curvilinear Flow Profiles Based on Reynolds Averaging. *J. Hydraul. Eng.* **2007**, *133*, 1074–1079. [CrossRef]
57. Kundu, S. Suspension concentration distribution in turbulent flows: An analytical study using fractional advection-diffusion equation. *Phys. A Stat. Mech. Its Appl.* **2017**, *506*, 135–155. [CrossRef]

58. Marcou, O.; Chopard, B.; Yacoubi, S.E.; Hamroun, B.; Lefevre, L.; Mendes, E. A Lattice Boltzmann Model to Study Sedimentation Phenomena in Irrigation Chanals. *Commun. Comput. Phys.* **2013**, *13*, 880–899. [CrossRef]
59. Dolanský, J.; Chára, Z.; Vlasák, P.; Kysela, B. Lattice Boltzmann Method used to Simulate Particle Motion in a Conduit. *J. Hydrol. Hydromech.* **2017**, *65*, 105–113. [CrossRef]
60. Shi, H.; Si, P.; Dong, P.; Yu, X. A Two-Phase SPH Model for Massive Sediment Motion in Free Surface Flows. *Adv. Water Resour.* **2019**, *129*, 80–98. [CrossRef]
61. Fourtakas, G.; Rogers, B.D. Modelling Multi-Phase Liquid-Sediment Scour and Resuspension Induced by Rapid Flows Using Smoothed Particle Hydrodynamics (SPH). *Accel. Graph. Process. Unit (GPU) Adv. Water Resour.* **2016**, *92*, 186–199. [CrossRef]

MDPI

Article

Self-Preservation of Turbulence Statistics in the Wall-Wake Flow of a Bed-Mounted Horizontal Pipe

Kalpana Devi [1], Prashanth Reddy Hanmaiahgari [1,*], Ram Balachandar [2] and Jaan H. Pu [3,*]

[1] Department of Civil Engineering, IIT Kharagpur, Kharagpur 721302, India; kalpanarajpoot@iitkgp.ac.in
[2] Department of Civil and Environmental Engineering, University of Windsor, Windsor, ON N9B 3P4, Canada; rambala@uwindsor.ca
[3] Faculty of Engineering and Informatics, School of Engineering, University of Bradford, Bradford BD7 1DP, UK
* Correspondence: hpr@civil.iitkgp.ac.in (P.R.H.); j.h.pu1@bradford.ac.uk (J.H.P.)

Abstract: This research article analyzed the self-preserving behaviour of wall-wake region of a circular pipe mounted horizontally over a flat rigid sand bed in a shallow flow in terms of mean velocity, RSS, and turbulence intensities. The study aims to investigate self-preservation using appropriate length and velocity scales.in addition to that wall-normal distributions of the third-order correlations along the streamwise direction in the wake region are analyzed. An ADV probe was used to record the three-dimensional instantaneous velocities for four different hydraulic and physical conditions corresponding to four cylinder Reynolds numbers. The results revealed that the streamwise velocity deficits, RSS deficits, and turbulence intensities deficits distributions displayed good collapse on a narrow band when they were non-dimensionalized by their respective maximum deficits. The wall-normal distance was non-dimensionalized by the half velocity profile width for velocity distributions, while the half RSS profile width was used in the case of the RSS deficits and turbulence intensities deficits distributions. The results indicate the self-preserving nature of streamwise velocity, RSS, and turbulence intensities in the wall-wake region of the pipe. The third-order correlations distributions indicate that sweep is the dominant bursting event in the near-bed zone. At the same time, ejection is the dominant bursting event in the region above the cylinder height.

Keywords: self-preservation in wall-wake; circular pipe; velocity deficit; RSS deficit; turbulence intensities deficit; third-order correlations

Citation: Devi, K.; Hanmaiahgari, P.R.; Balachandar, R.; Pu, J.H. Self-Preservation of Turbulence Statistics in the Wall-Wake Flow of a Bed-Mounted Horizontal Pipe. *Fluids* **2021**, *6*, 453. https://doi.org/10.3390/fluids6120453

Academic Editor: Rob Poole

Received: 17 September 2021
Accepted: 23 November 2021
Published: 14 December 2021

1. Introduction

The wake phenomenon is widespread in the nature, and its prevalent examples are environmental flows and flows occurring in many engineering applications. The wake is the highly turbulent region behind the bodies laid on the flat surfaces formed due to flow separation. The turbulent properties of the cylinder wake are a well-researched topic in fluid dynamics. Many investigations are available in the past literature [1–7] to study the cylinder wake. Djeridi et al. [1] conducted experimental research to explore the turbulent characteristics in the flow field of a circular cylinder in the near-bed upstream and near-wake region for a high value of Reynolds number equal to 14,000. They also analyzed the mean and turbulent flow field for a moderate blockage and an aspect ratio. Konstantinidis et al. [2] experimentally determined the mean and fluctuating velocity fields in the near wake of a circular cylinder under periodic velocity fluctuations incident on the mean flow. They described the wake in terms of the recirculation region, vortex formation region, the maximum intensity of velocity fluctuations, and vortex street's wavelength. Braza et al. [3] studied the flow field of a circular cylinder with a high blockage and a low aspect ratio. The flow field was explored in the starting point of the critical flow regime for a Reynolds number of 14,000. Akoz [4] experimentally estimated the characteristics of flow structures behind the cylinder, instantaneous and time-averaged velocity field, vorticity contours, streamline topology, and Reynolds stress concentrations for the cylinder Reynolds number, Re_D between 1000 and 7000. They dictated separation point from the cylinder surface and the variation of length of the primary and secondary downstream separation regions with Reynolds numbers. Akoz and Kirkgoz [5] investigated the turbulent flow field of a bed-mounted horizontal cylinder for Reynolds numbers between 1000 and 7000. Devi and Hanmaiahgari [6] experimentally analyzed a wall-mounted cylinder's mean and turbulent characteristics at a downstream location. Devi et al. [7] conducted a comparative study for the flow field of a wall-mounted cylinder to determine the effect of gravel-bed roughness compared to sand bed roughness. Despite numerous research in this field, many issues related to the cylinder wake are still unresolved, especially in shallow flows.

In the shallow flows, the horizontal length scale is generally larger than the wall-normal length scale [8]. Destabilizing shallow flows are caused by an abrupt change in topology, such as a bluff body over a flat surface, resulted in shallow wakes. Many researchers [9–11] have studied the nature of shallow wakes. Akilli and Rockwell [11] used flow visualization techniques in addition to particle image velocimetry (PIV) to study the shallow flows in the near wake region of a circular cylinder. They discussed the variation of mean velocity, Reynolds stress, vorticity, and streamline topology at three different wall-normal distances. The channel bed can significantly affect the flow characteristics of the shallow wake, and bed effects are attenuated while moving away from it towards the free surface. In the near-bed region of the flow, the bed restricts the development of the wake, diminishes the interaction between shear layers, and stabilizes the flow [12].

Self-preservation of turbulent wakes downstream of a cylinder at high Reynolds numbers was initially reported by Townsend [13,14] in his elaborate investigations. Out of these wakes, the one which frequently occurs in the environment is the shallow wake. The concept of self-preservation of turbulent flows is also known as local similarity or self-similarity. If some or all statistical characteristics of a turbulent flow are a function of only scales, the flow is called self-preserving in its behaviour. In other words, a turbulent flow in its self-preserving condition is the only function of its length and velocity scales. The far-field of a turbulent free shear layer such as a wake and a jet are classic examples of self-preserving flows. Maji et al. [15] reviewed the occurrence of von Kármán vortex streets as a function of solid volume fraction in the wake region of a group of wall-mounted vertical cylinders. Sadeque et al. [16] analyzed the flow patterns of the near-wakes of the wall-mounted vertical cylinder in the turbulent shallow flow. They found that flow fields in the region away from the wall are similar. They also perceived that the turbulent kinetic energy and the primary Reynolds stress distributions displayed the self-preserving behaviour in moderate to deeply submerged cylinders. Balachandar et al. [8] and Tachie and Balachandar [17] deduced that even in a near-wake region of the flow, the self-preserving characteristics can be established in terms of the time-averaged velocity distributions by using appropriate velocity and length scales. Maji et al. [18] analyzed the self-preserving behaviour in the interior and the wake region of the vegetation patch regarding the mean flow velocities in the streamwise and wall-normal directions and turbulence intensities, RSS, and turbulent kinetic energy (TKE). They observed that the time-averaged streamwise velocity distributions depict self-preserving characteristics for both emergent and sparse vegetation patches in their outer, interior, and wake regions when scaled by their depth-averaged value. The self-preserving behaviour of the time-averaged streamwise velocity distributions was independent of the aspect ratios and the Reynolds numbers. Similar observations were also made for the RSS profiles. In the past literature [8,16,17], different velocity and length scales were used for velocity, RSS, and turbulence intensities distributions. Balachandar et al. [8] and Tachie and Balachandar [17] used $U_{max} - U_{min}$ as a velocity scale and b as a length scale for the velocity distributions to establish the self-preserving characteristics of the wake region of a sharp-edged plate. Here, U_{max} and U_{min} are the maximum and the minimum velocities, respectively, in the wake region at a particular streamwise distance, x/D, and b is the wall-normal distance at which $U - U_{min}$ is equal to one half of the $U_{max} - U_{min}$. Sadeque et al. [16] used U_{1m} as the velocity scale for velocity profiles for the bed-mounted cylinder. In their study, the U_{1m} is equivalent to

$U_{max} - U_{min}$ of Balachandar et al. [8] and Tachie and Balachandar [17]. They also observed an excellent collapse of the velocity profiles in the near wake region of the cylinder. The velocity deficit profiles collapse on a narrow band for this scaling, representing that the wall-wake region of the pipe is self-preserving in nature. Despite that much research in this area, there are very few studies that explore the self-preserving characteristics of the wall-wake region of a bed-mounted horizontal circular pipe in the shallow flow with proper scaling for the velocity, RSS, and turbulence intensities.

Therefore, the main objective of this study is to assess the self-preserving characteristics of the wall-wake flow of a bed-mounted circular pipe in terms of streamwise velocity, RSS, and turbulence intensities for different inflow conditions. Further, the variation of the wall-normal distributions of third-order correlations along the streamwise direction in the wake region was also evaluated. For this purpose, an ADV probe was used to measure instantaneous three-dimensional (3D) velocity data. After post-processing the velocity data, it was used to determine turbulence statistics of the flow and their deficits and maximum values. Henceforth, how the wall-normal profiles of the time-averaged streamwise velocity u^+, RSS ${\tau_{uw}}^+$, streamwise turbulence intensity ${\sigma_u}^+$, wall-normal turbulence intensity ${\sigma_w}^+$ in non-dimensional forms and third-order correlations M_{jk} change in the wake region of the pipe with streamwise distance, $\hat{x}$ were also documented. Additionally, the wall-normal distributions of non-dimensional velocity deficit $\Delta u/(\Delta u)_{max}$, non-dimensional RSS deficit $\Delta\tau_{uw}/(\Delta\tau_{uw})_{max}$, non-dimensional streamwise turbulence intensity deficit $\Delta\sigma_u/(\Delta\sigma_u)_{max}$ and non-dimensional wall-normal turbulence intensity deficit $\Delta\sigma_w/(\Delta\sigma_w)_{max}$ were investigated to determine the self-preserving behaviour of velocity, RSS, and turbulence intensities, respectively. Besides this, the decay of maximum velocity deficit $(\Delta u)_{max}/u_*$, maximum RSS deficit $-(\Delta\tau_{uw})_{max}/{u_*}^2$, maximum streamwise turbulence intensities deficit $-(\Delta\sigma_u)_{max}/u_*$ and maximum wall-normal turbulence intensities deficit $-(\Delta\sigma_w)_{max}/u_*$ with streamwise distance, $\hat{x}$ were also evaluated. This was followed by the analysis of the streamwise growth of half velocity profile width, z_1 and half RSS profile width, z_2.

2. Experimental Setup and Data Measurement Procedure

This section presents the experimental channel description, sand bed material analysis, pipe model details, ADV probe specifications, a summary of the hydraulic and physical conditions of experimental runs, data measurement procedure, and data processing techniques used.

2.1. Experimental Setup

A rectangular closed-loop channel of 12 m length with a cross-section of $B \times H = 0.91$ m $\times$ 0.70 m was utilized for the experimental purpose [6,7]. A constant streamwise bed slope of 0.23% was maintained along the channel length for all the experimental runs. The channel side walls at the test section consisted of transparent glass to facilitate visualization. At the channel inlet, two flow straighteners were provided to suppress the pump vibrations. A circular pipe of diameter D was laid on the bed at a distance of 6.5 m downstream of the channel inlet. The pipe was laid in such a way that it was not allowing the flow underneath it. Uniform sand having a median size, d_{50} equal to 2.54 mm, was coated over the bed to generate the rough surface. A user-configured tailgate was provided at the channel outlet to regulate the flow depth. The water surface levels were measured by employing two sets of Vernier point gauges with an accuracy of ± 0.1 mm. The ADV system was carried on a movable trolley supported on two rails over the channel walls.

The streamwise direction of the coordinate system used in the experiment is taken as the x-axis with the positive in the downstream direction and the negative in the upstream direction. The transverse direction is taken as the y-axis with the positive towards the left of the channel centerline and the negative on the right side of the central plane of the channel. The z-axis was oriented in the wall-normal direction to the channel bed with a positive in the vertically upward direction. The origin of the coordinate system was located at the intersection point of the axis of the pipe and the central wall-normal plane (xz).

A four-receivers down-looking ADV probe (Vectrino plus) manufactured by Nortek company (Nortek AS, Vangkroken 2, NO-1351 RUD, Norway), operating with an acoustic frequency of 10 MHz, was utilized to record 3D instantaneous velocity data. The Vectrino collect the velocity data at a location 5 cm below the transmitting probe. The sampling volume cylinder diameter and height both are 6 mm. The data were recorded for a sampling time equal to five minutes with a sampling frequency of 100 Hz. The sampling time of five minutes was found to be adequate for obtaining self-determining time-averaged flow velocities [19]. Unwanted spikes mostly corrupted the near-bed data; therefore, despiking was adopted to separate the noise from the raw data using the despiking techniques suggested by previous researchers [20–22], and replaced using a cubic interpolation method. To validate the quality of the data obtained after despiking, the signal-to-noise ratio (SNR) test as proposed by Chanson et al. [23] was implemented. The data threshold values of SNR and correlation coefficient as 17 and 70%, respectively, were maintained. The uncertainty statistics of the ADV probe were analyzed by collecting 10 samples at 10 mm above the bed level, for a sampling duration of 5 min at a sampling rate of 100 Hz. The uncertainties in the velocity terms, Reynolds stresses and turbulence intensities terms, and third-order correlations are less than 5%, 10%, and 13%, respectively.

2.2. Data Measurement Procedure

The channel was left unused for an hour to stabilize the flow after starting the pump. All data measurements were taken in the central wall-normal plane (xz) of the channel. Four experimental runs, namely, Run 1, Run 2, Run 3, and Run 4, were conducted corresponding to four different cylinder Reynolds numbers (Re_D). The summary of all four experimental runs is given in Table 1.

Table 1. The summary of all four experimental runs.

Exp. Run	D (m)	h (m)	U (m/s)	u_* (m/s)	F_r	Re_D
Run 1	0.05	0.3	0.15	0.0071	0.111	7500
Run 2	0.05	0.3	0.19	0.0092	0.111	9500
Run 3	0.08	0.3	0.15	0.0071	0.087	12,000
Run 4	0.08	0.3	0.19	0.0092	0.111	15,200

Two different pipe diameters, D of 5 and 8 cm, were used as pipe models. A constant flow depth, h of 0.3 m, was maintained for all four runs. Two approach flow area-averaged velocities, U of 0.15 and 0.19 m/s, were maintained during the complete experiment. The cylinder Reynolds number and the Froude number were estimated as $Re_D = UD/\nu$, and $F_r = U/\sqrt{gh}$, respectively. All experimental runs were conducted under subcritical flow conditions, as is clear from the value of F_r from Table 1. The approach flow shear velocity, u_* was determined by linearly extending the RSS profile to the channel bed [24].

The data were recorded at nine streamwise locations ($\hat{x}$ = −4, 0, 2, 3, 4, 6, 8, 10, and 12) for each run. Here, $\hat{x}$ is the non-dimensional streamwise distance, taken as the ratio of streamwise distance, x to the pipe diameter, D and written as $\hat{x} = x/D$. The $\hat{x}$ = −4 is an undisturbed upstream location, $\hat{x}$ = 0 is located just above the pipe, and the rest of the data measurement locations are in the wake region of the pipe. For each measurement location, data were collected at 26 wall-normal points as z = 0.003, 0.005, 0.007, 0.009, 0.012, 0.015, 0.02, 0.025, 0.03, 0.035, 0.04, 0.045, 0.05, 0.06, 0.07, 0.08, 0.09, 0.1, 0.11, 0.13, 0.15, 0.17, 0.19, 0.21, 0.23 and 0.245 m except at $\hat{x}$ = 0. At $\hat{x}$ = 0, it was not feasible to record data below the top level of pipe, so z = 0.06 and 0.09 m onwards were collected for 5 cm and 8 cm diameter pipes, respectively. A schematic diagram of the experimental setup showing its plane view with the streamwise measurement locations for Run 1, Run 2, Run 3, and Run 4, along with the coordinate system used for the experiments, is shown in Figure 1.

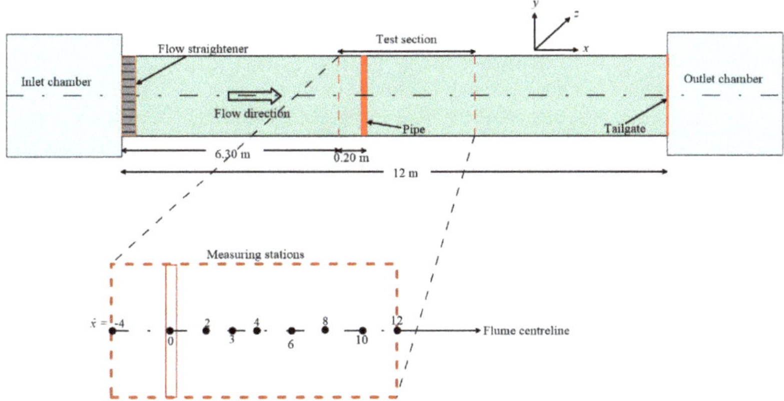

Figure 1. Schematic diagram of the experimental setup showing its plane view with the streamwise measurement locations for Run 1, Run 2, Run 3, and Run 4, and the coordinate system used for the experiments.

Figure 2 depicts the vector plots of the non-dimensional resultant velocity, U^+ along the non-dimensional central wall-normal plane, xz ($\hat{x}$, z/D) of the channel for Run 3 in the wall-wake region of the pipe. The diameter of pipe D was used as the streamwise length scale to non-dimensionalize streamwise distance, x as $\hat{x}(= x/D)$. The vector plots were shown only up to a wall-normal domain of $2D$ to depict the flow field in the near-bed region. The approach flow shear velocity, u_* is used as velocity scale to normalize the velocity vector of magnitude, $\overline{U} = \sqrt{\overline{u}^2 + \overline{w}^2}$ with the direction, $\theta = \tan^{-1}(\overline{w}/\overline{u})$. Here, θ is the angle made by the velocity vector with the horizontal direction. It was observed from Figure 2 that the flow separated at the top surface of the pipe and later reattached to the channel bed downstream with the formation of a recirculation region in the near-bed flow zone. The recirculation region is stretched up to $\hat{x} = 7.4$. Velocity vectors are pointed in the opposite direction of the flow in the recirculation region's near-bed zone, indicating the existence of the resultant negative velocity in the region. The dotted blue line demarcates the recirculation region in Figure 2. Beyond this line, the mean velocity becomes entirely positive. After the reattachment point, the flow starts becoming similar to its undisturbed upstream form. It is noteworthy that within the streamwise domain of $12D$, the mean velocity is not fully recovered to its undisturbed upstream value.

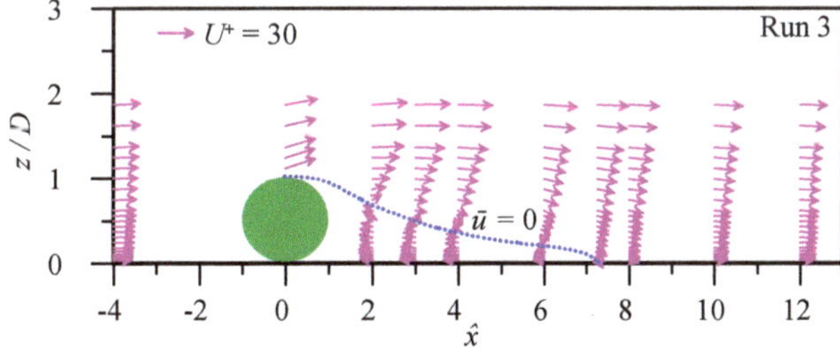

Figure 2. Vector plot of non-dimensional resultant velocity, U^+ along the central wall-normal plane (xz) of the channel for Run 3 in the wall-wake region of the pipe.

3. Results and Discussion

The approach flow shear velocity, u_*, was used as the velocity scale to non-dimensionalize the turbulent characteristics and the flow depth, h, was used as the wall-normal length scale for the wall-normal distance, z, as $\bar{z}$ $(= z/h)$. The wall-normal profiles of the non-dimensional time-averaged streamwise velocity, $u^+(= \bar{u}/u_*)$, in the wake region of the pipe at a different streamwise distance, $\hat{x}$, for Run 3 and Run 4 are plotted in Figure 3. It is clear from Figure 3 that the streamwise velocity, u^+, has a negative sign in the near-bed region because of flow separation while it becomes positive in the region away from the bed. The velocity in the free-surface region is also affected by the presence of the pipe, and its magnitude is different from the corresponding approach flow velocity in the free surface region. The velocity in the free-surface region of the recirculation region is greater than the undisturbed upstream velocity at the same elevation because of lower near-bed velocities in the recirculation region which shows mass conservation. In addition, this result reveals that the flow was diverted upward and accelerated in the near-surface region due to the separation of the flow from the pipe. The maximum magnitude of the streamwise velocity is occurring near the free surface for all streamwise locations, $\hat{x}$. The maximum magnitude of the streamwise velocity is increasing along with streamwise distance up to $\hat{x} = 6$ and then decreasing with a further increase in $\hat{x}$. It is noteworthy that the negative velocity is fully recovered at a streamwise distance of $\hat{x} = 8$, in both the runs (Run 3 and Run 4). The velocity, u^+-, distributions in the wake region display an upward concavity in their profiles, and this asymmetric velocity distribution occurs due to the wall-wake effects.

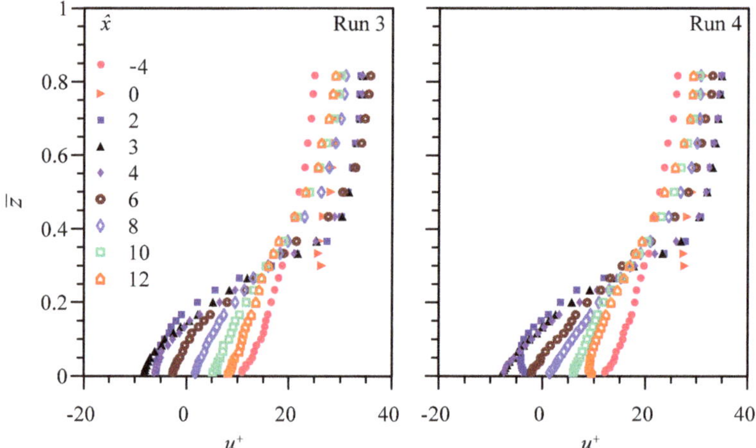

Figure 3. Wall-normal profiles of non-dimensional time-averaged streamwise velocity, u^+, in the wake region of the pipe at different streamwise locations, $\hat{x}$, for Run 3 and Run 4.

The points of inflection ($d^2\bar{u}/dz^2 = 0$) in the individual u^+ profiles are observed near the top level of the pipe, $z/D \approx 1$. In the free-surface region, the velocity u^+- distributions are attaining almost the same velocity gradient, $d\bar{u}/dz$. The extrapolation of the streamwise velocity profiles up to the channel bed verified the preservation of the no-slip condition.

Figure 4 depicts the schematic diagram of a typical wall-normal distribution of streamwise velocity deficit, Δu, in the wall-wake region of the pipe along with the definition of the half velocity profile width, z_1. The velocity deficit, Δu, is the difference between the upstream undisturbed time-averaged streamwise velocity at particular elevation z and the downstream wall-wake time-averaged streamwise velocity at the same elevation. The velocity deficit, Δu (z), at a specific wall-normal distance, z, can be written as $\Delta u\ (z) = u_{u/s}(z) - u_{d/s}(z)$. Here, $u_{u/s}(z)$ is the upstream undisturbed time-averaged streamwise velocity at an elevation, z, while $u_{d/s}(z)$ is the downstream wall-wake time-averaged streamwise velocity at the same elevation. The half velocity profile width, z_1, is the wall-normal elevation from the bed at which the velocity deficit is equal to one-half of the maximum velocity deficit, $(\Delta u)_{max}$.

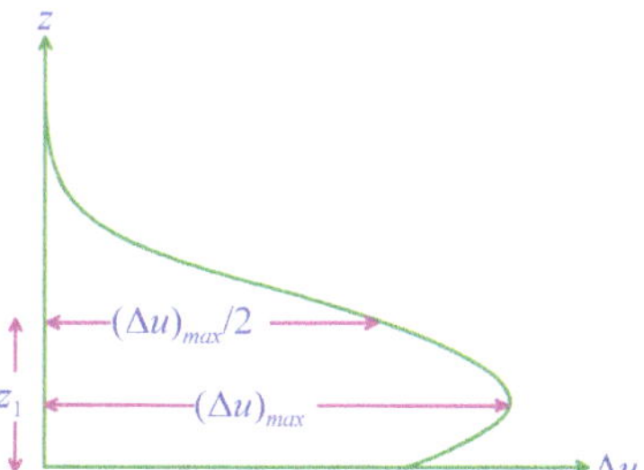

Figure 4. Schematic representation of a typical wall-normal distribution of streamwise velocity deficit, Δu, in the wall-wake region of the pipe along with the definition of the half velocity profile width, z_1.

Figure 5a displays the variation of $\Delta u/(\Delta u)_{max}$ against z/z_1 for Run 1, Run 2, Run 3, and Run 4 in the wall-wake region of the pipe, depicting the self-preserving characteristics of the velocity deficit, Δu, distributions. The maximum velocity deficit, $(\Delta u)_{max}$, is used as the velocity scale, and the half velocity profile width, z_1, is taken as the wall-normal length scale. The velocity deficit profile is exhibiting a single peak in its distribution at $z/z_1 \approx 0.25$. The self-preserving characteristics of the velocity deficit profiles are found to be independent of approach flow conditions and diameter of the cylinder, D. All of the streamwise velocity profiles are showing a good collapse over a narrow band for the wall-normal domain of the study, $z/D < 2$, and for the complete streamwise domain of the research. Above the wall-normal domain, $z/D > 2$ velocity profiles are not self-preserving in nature; therefore, they are excluded from the analysis. It can also be observed that the value of non-dimensional velocity deficit is equal to unity at $z/z_1 \approx 0.25$, which simply implied that $z \approx 0.25z_1$ is the wall-normal location of the maximum velocity deficit, $(\Delta u)_{max}$.

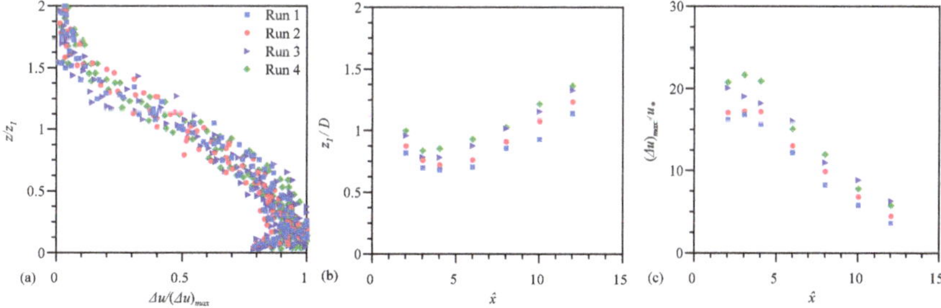

Figure 5. Self-preservation analysis of streamwise velocity distribution in the wall-wake region of the pipe for Run 1, Run 2, Run 3, and Run 4: (**a**) variation of $\Delta u/(\Delta u)_{max}$ against z/z_1, depicting the self-preservation in velocity deficit (Δu) distribution; (**b**) streamwise growth of non-dimensional half velocity profile width, z_1/D; and (**c**) streamwise decay of non-dimensional maximum velocity deficit, $(\Delta u)_{max}/u_*$.

Figure 5b portrays the streamwise growth of non-dimensional half velocity profile width, z_1/D for Run 1, Run 2, Run 3, and Run 4 in the wall-wake region of the pipe. The diameter of pipe D is used as the wall-normal length scale. It can be observed from Figure 5b that the half velocity profile width, z_1/D, grows consistently and gradually except for starting two points. This variation in the half velocity profile width is resulting from an increase followed by a decrease in the thickness of the wake region downstream of the cylinder. A similar observation was also made by Tachie and Balachandar [17], with a slightly different value of streamwise distance as $\hat{x} \geq 5$, the half velocity profile width grows constantly and gradually with $\hat{x}$. Contrary to this, Sadeque et al. [16] found a constant decline in the half velocity profile width with streamwise distance for a bed-mounted vertical cylinder.

The streamwise decay of non-dimensional maximum velocity deficit, $(\Delta u)_{max}/u_*$, for Run 1, Run 2, Run 3, and Run 4 are demonstrated in Figure 5c. The approach flow shear velocity, u_*, is taken as the velocity scale to normalize the maximum velocity deficit, $(\Delta u)_{max}$. It is perceived from Figure 5c that the maximum velocity deficit, $(\Delta u)_{max}/u_*$, decays in the streamwise direction with $\hat{x}$. Its decay indicates the recovery of the wall-wake flow velocity similar to its undisturbed upstream values. The flow is uniformly and quickly recovering to its undisturbed upstream values. Schmeeckle and Nelson [16] also made similar observations.

The wall-normal profiles of the non-dimensional RSS, ${\tau_{uw}}^+$ in the wake region of the pipe at a different streamwise distance, $\hat{x}$, for Run 3 and Run 4 are shown in Figure 6. Here, Reynolds shear stress (RSS), τ_{uw} is expressed as $\tau_{uw} = -\overline{u'w'}$. It is visible from Figure 6 that RSS starts with a small positive value at the bed and increases sharply with an increase in the wall-normal distance, z. It attained its peak at the top level of the pipe ($z/D \approx 1$) and decreased quickly with a further rise in $\overline{z}$. Henceforth, the value of RSS becomes very small and almost invariant of the wall-normal distance, z in the free-surface region. The location of the peak of the RSS corresponds to the point of inflection ($d^2\overline{u}/dz^2 = 0$) in the velocity profiles. It is understood that the peaks in RSS at the top level of the pipe are caused by the enhanced mixing of turbulence at that level.

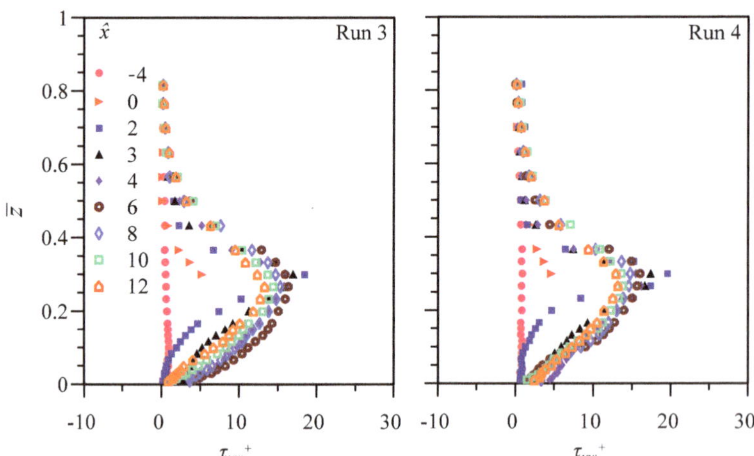

Figure 6. Wall-normal profiles of non-dimensional RSS, ${\tau_{uw}}^+$, in the wake region of the pipe at different streamwise locations, $\hat{x}$, for Run 3 and Run 4.

Figure 7 illustrates the diagram of the typical wall-normal distribution of RSS deficit, $\Delta\tau_{uw}$, in the wall-wake region of the pipe along with the definition of the half RSS profile width, z_2. The RSS deficit, $\Delta\tau_{uw}$, is the difference between the downstream wall-wake RSS at an elevation, z, and upstream undisturbed RSS at the same elevation. The RSS deficit, $\Delta\tau_{uw}(z)$, at a particular elevation, z, can be written as $\Delta\tau_{uw}\ (z) = \tau_{u/s}(z) - \tau_{d/s}(z)$. Here, $\tau_{u/s}(z)$ is the upstream undisturbed RSS at an elevation, z, and $\tau_{d/s}(z)$ is the downstream wall-wake RSS at the same elevation. The half RSS profile width, z_2, was estimated similarly to the half velocity profile width, z_1. The half RSS profile width z_2 is the wall-normal height from the bed at which the RSS deficit is equal to $0.5(\Delta\tau_{uw})_{max}$.

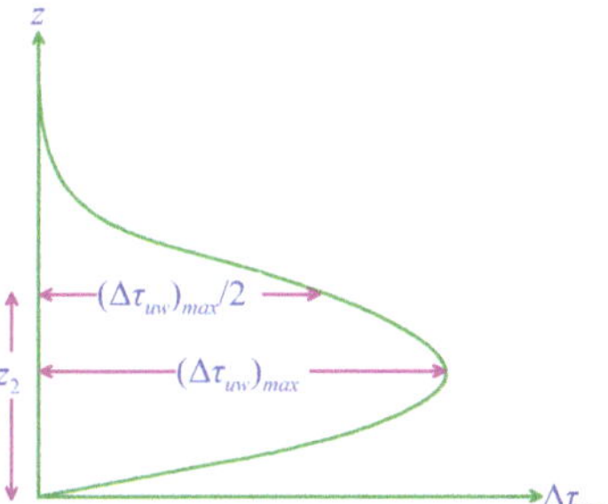

Figure 7. Schematic representation of a typical wall-normal distribution of RSS deficit, $\Delta\tau_{uw}$, distribution in the wall-wake region of the pipe along with the definition of the half RSS profile width, z_2.

Figure 8a expresses the variation of $\Delta\tau_{uw}/(\Delta\tau_{uw})_{max}$ against z/z_2 for Run 1, Run 2, Run 3, and Run 4 in the wall-wake region of the pipe, by depicting the self-preserving characteristics in the RSS deficit (Δu) profiles. The self-preservation of the wall-wake is evaluated based upon the collapse of the RSS deficit profiles on a narrow band. The maximum RSS deficit, $(\Delta\tau_{uw})_{max}$ is used as velocity scale and the half RSS profile width, z_2 is taken as the wall-normal length scale. The RSS deficit profiles are collapsing on a narrow band for this scaling, which demonstrates that the wall-wake region of the pipe is self-preserving in nature. The self-preserving characteristics of the RSS deficit profiles are found to be independent of the approach flow conditions similar to the velocity deficit profiles. Similar to velocity profiles, all RSS profiles show good collapse with the streamwise distance, $\hat{x}$, for the wall-normal domain of study, $z/D < 2$. It is noteworthy that the RSS deficit profiles display a single peak, and it occurs at $z/z_1 \approx 0.6$. It can also be observed from Figure 8a that the value of non-dimensional RSS deficit $(\Delta\tau_{uw})_{max}$ was equal to unity for $z/z_1 \approx 0.6$. It simply reveals the peak value of the RSS deficit occurs at $z \approx 0.6z_1$ i.e., $\Delta\tau_{uw}|_{z=0.6z_1} = (\Delta\tau_{uw})_{max}$.

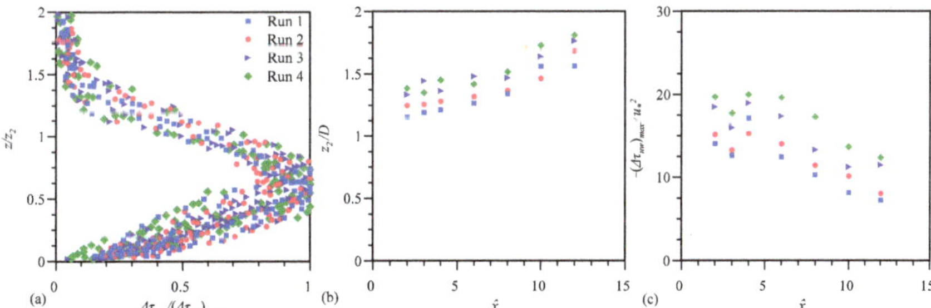

Figure 8. Self- preservation analysis of RSS distribution in the wall-wake region of the pipe for Run 1, Run 2, Run 3, and Run 4: (**a**) variation of $\Delta\tau_{uw}/(\Delta\tau_{uw})_{max}$ against z/z_2, depicting the self-preservation in RSS deficit ($\Delta\tau_{uw}$) distribution; (**b**) streamwise growth of non-dimensional half RSS profile width, z_2/D; and (**c**) streamwise decay of non-dimensional maximum RSS deficit,$-(\Delta\tau_{uw})_{max}/{u_*}^2$.

Figure 8b displays the streamwise growth of non-dimensional half RSS profile width, z_2/D, for Run 1, Run 2, Run 3, and Run 4 in the wall-wake region of the pipe. The diameter of the pipe, D, is used as the wall-normal length scale. It can be observed from Figure 8b that half RSS profile width, z_2/D, grows in the streamwise direction with an increase in the streamwise distance, $\hat{x}$.

The streamwise decay of non-dimensional maximum RSS deficit, $-(\Delta\tau_{uw})_{max}/u_*^2$ for Run 1, Run 2, Run 3, and Run 4 are depicted in Figure 8c. The negative sign is used before the maximum RSS deficit $-(\Delta\tau_{uw})_{max}/u_*^2$, since $\tau_{d/s}(z)$ is greater than $\tau_{u/s}(z)$. The approach flow shear velocity, u_*, is taken as the velocity scale to normalized the maximum RSS deficit, $-(\Delta\tau_{uw})_{max}$. Figure 8c demonstrates that the maximum RSS deficit, $-(\Delta\tau_{uw})_{max}/u_*^2$, decays in the streamwise direction with $\hat{x}$, which indicates the recovery of the flow with an increase in $\hat{x}$. Similar results were also revealed by Sadeque et al. [16], although their case is of a bed-mounted vertical cylinder. These patterns in the variation of $-(\Delta\tau_{uw})_{max}/u_*^2$ in the streamwise direction are happening due to the dampening of the turbulence mixing with an increase in $\hat{x}$ and recovering the flow similar to its undisturbed upstream form.

The turbulence intensity downstream of the pipe exists due to the fluctuations in the instantaneous flow velocity influenced by the pipe and bed roughness. Non-dimensional turbulence intensities in the streamwise, $\sigma_u{}^+$ and wall-normal directions, $\sigma_w{}^+$ are given as $\sigma_u{}^+ = \sigma_u/u_*$ and $\sigma_w{}^+ = \sigma_w/u_*$, respectively. The streamwise, $\sigma_u \left(= \sqrt{\overline{u'u'}}\right)$ and wall-normal, σ_w $(= \sqrt{\overline{w'w'}})$ turbulence intensities are the root mean square (RMS) of the fluctuating velocity components in respective directions, representing the strength of turbulence. The wall-normal profiles of the non-dimensional streamwise, $\sigma_u{}^+$ and wall-normal turbulence intensities, $\sigma_w{}^+$ in the wake region of the pipe at a different streamwise distance, $\hat{x}$ for Run 3 and Run 4 are depicted in Figures 9 and 10. It is observed from Figures 9 and 10 that the wall-normal distributions of turbulence intensities are akin to RSS distribution with smaller magnitudes. Similar to RSS distributions, at the bed, $\sigma_u{}^+$ and $\sigma_w{}^+$ for both runs, are starting with small positive values, and increasing sharply with an increase in $\bar{z}$ until they attain their peaks at the top-level of the pipe ($z/D \approx 1$), which indicates the corresponding high-velocity fluctuations due to enhanced turbulence mixing at that level. Turbulence intensities are declining with a further rise in $\bar{z}$ for $z/D > 1$. The peaks of turbulence intensities show increasing trends in the near-wake region and decreasing patterns in the far-wake region along the streamwise direction, $\hat{x}$. These patterns are attributed to the enhanced turbulence in the former region and the attenuated turbulence in the latter region. Below the top level of the pipe, $z/D \leq 1$, the turbulence intensities dampened with a fall in $\bar{z}$ due to a reduction in the turbulence mixing with a decline in $\bar{z}$. In the free-surface region, $\bar{z} > 0.8$, turbulence intensities attained very small magnitudes and became approximately constant against $\bar{z}$. This result indicates that the free-surface region is far away from the influence of the bed and the cylinder.

Figure 11a,b display the variation of $\Delta\sigma_u/(\Delta\sigma_u)_{max}$ and $\Delta\sigma_w/(\Delta\sigma_w)_{max}$, respectively, against z/z_2 for Run 1, Run 2, Run 3, and Run 4 in the wall-wake region of the circular pipe, depicting the self-preserving characteristics of the turbulence intensities distributions. The self-preservation of the wall-wake is also evaluated based upon the collapse of the turbulence intensities deficit profiles. The maximum turbulence intensities deficits, $(\Delta\sigma_u)_{max}$ and $(\Delta\sigma_w)_{max}$, are used as velocity scales, respectively, for $\Delta\sigma_u$ and $\Delta\sigma_w$, and the half RSS profile width, z_2, is taken as the wall-normal length scale. The turbulence intensities deficits, $\Delta\sigma_u$ and $\Delta\sigma_w$, are calculated as the difference between the downstream wall-wake turbulence intensities at a particular elevation, z, and upstream undisturbed turbulence intensities at the same elevation. The turbulence intensities deficits, $\Delta\sigma_u(z)$ and $\Delta\sigma_w(z)$, at a particular elevation, z, can be written as $\Delta\sigma_u\ (z) = \sigma_u(z)_{d/s} - \sigma_u(z)_{u/s}$ and $\Delta\sigma_u\ (z) = \sigma_u(z)_{d/s} - \sigma_u(z)_{u/s}$, respectively. Here, $\sigma_u(z)_{u/s}$ and $\sigma_w(z)_{u/s}$ are the upstream undisturbed turbulence intensities in streamwise and wall-normal directions, respectively, at an elevation, z, while $\sigma_u(z)_{d/s}$ and $\sigma_w(z)_{d/s}$ are turbulence intensities in streamwise and wall-normal directions, respectively, in the wall-wake region of the pipe at particular $\hat{x}$ at the same elevation. The turbulence intensities deficit profiles collapse on a narrow band when scaled by these characteristic scales of velocity and length, representing that the wall-wake region of the pipe is self-preserving in nature. The self-preserving behaviour of the turbulence intensities distributions is found to be independent of approach flow conditions. All of the turbulence intensities profiles show good collapse with the streamwise distance, $\hat{x}$, for the wall-normal domain of the study, $z/D < 2$.

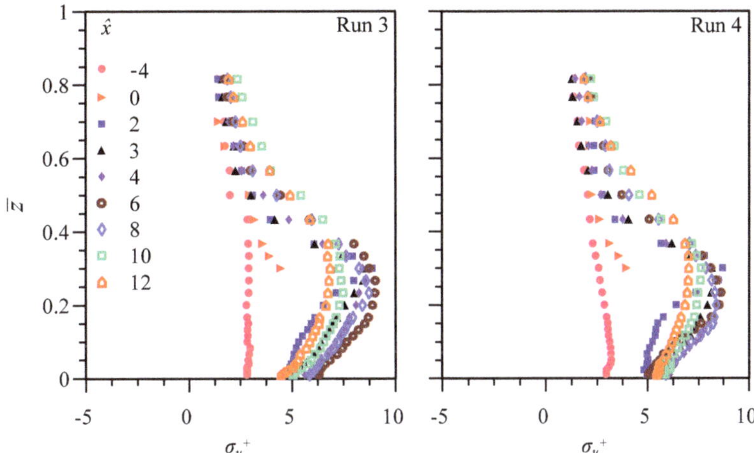

Figure 9. Wall-normal profiles of non-dimensional streamwise turbulence intensity, $\sigma_u{}^+$ in the wake region of the pipe at different streamwise locations, $\hat{x}$, for Run 3 and Run 4.

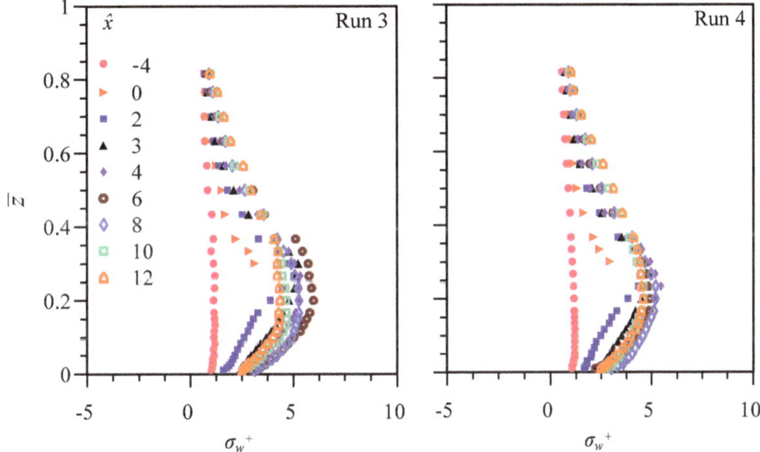

Figure 10. Wall-normal profiles of non-dimensional wall-normal turbulence intensity, $\sigma_w{}^+$, in the wake region of the pipe at different streamwise locations, $\hat{x}$, for Run 3 and Run 4.

The streamwise decay of non-dimensional maximum turbulence intensities deficits, $-(\Delta\sigma_u)_{max}/u_*$ and $-(\Delta\sigma_w)_{max}/u_*$ for Run 1, Run 2, Run 3, and Run 4 are illustrated in Figure 11c,d. The approach flow shear velocity, u_*, is taken as the velocity scale to normalize the maximum turbulence intensities deficits, $-(\Delta\sigma_u)_{max}$ and $-(\Delta\sigma_w)_{max}$. It is observed from Figure 11c that the maximum streamwise turbulence intensity deficit decreases slowly with $\hat{x}$. However, the maximum wall-normal turbulence intensity deficit increases in the beginning and attaining maximum value at $\hat{x} = 6$, and after that, it falls with a further rise in the streamwise distance, as is visible from Figure 11d.

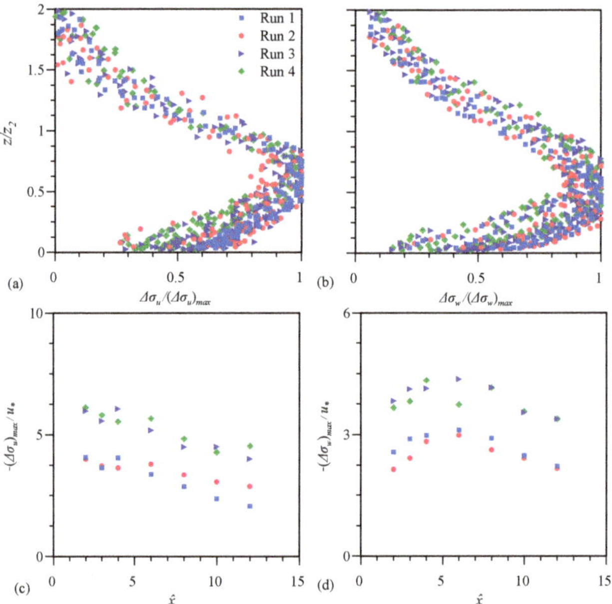

Figure 11. Self- preservation analysis of turbulence intensities in the wall-wake region of the circular pipe for Run 1, Run 2, Run 3, and Run 4: (**a**) variation of $\Delta\sigma_u/(\Delta\sigma_u)_{max}$ against z/z_2, depicting the self-preservation in streamwise turbulence intensity deficit ($\Delta\sigma_u$) distribution; (**b**) variation of $\Delta\sigma_w/(\Delta\sigma_w)_{max}$ against z/z_2, depicting the self-preservation in wall-normal turbulence intensity deficit ($\Delta\sigma_w$) distribution; (**c**) streamwise decay of non-dimensional maximum streamwise turbulence intensity deficit,$-(\Delta\sigma_u)_{max}/u_*$; (**d**) streamwise decay of non-dimensional maximum wall-normal turbulence intensity deficit,$-(\Delta\sigma_w)_{max}/u_*$.

The third-order correlations give valuable information about the fluxes and diffusion of the Reynolds normal stresses (RNSs) [25]. To be precise, the effective response of the bursting events can also be determined by the third-order correlations [24]. The third-order correlations (M_{jk}) are expressed as $M_{jk} = \overline{\hat{u}'^j\hat{w}'^k}$ where $j + k = 3$, $\hat{u} = u'/\left(\overline{u'u'}\right)^{0.5}$, and $\hat{w} = w'/\left(\overline{w'w'}\right)^{0.5}$. The third-order correlations can be categorized into four categories, namely: flux of streamwise RNS in the flow direction, $M_{30} = \overline{\hat{u}^3} = \overline{u'^3}/\left(\overline{u'u'}\right)^{3/2}$; flux of the wall-normal RNS in the wall-normal direction, $M_{03} = \overline{\hat{w}^3} = \overline{w'^3}/\left(\overline{w'w'}\right)^{3/2}$; diffusion of streamwise RNS in wall-normal direction, $M_{21} = \overline{\hat{u}^2\hat{w}} = \overline{u'^2w'}/\left(\overline{u'u'}\right)\left(\overline{w'w'}\right)^{1/2}$; and diffusion of wall-normal RNS in the streamwise direction, $M_{12} = \overline{\hat{u}\hat{w}^2} = \overline{u'w'^2}/\left(\overline{u'u'}\right)^{1/2}\left(\overline{w'w'}\right)$. The wall-normal profiles of the third-order correlations, M_{jk}, in the wake region of the pipe at different $\hat{x}$ for Run 3 and Run 4 are illustrated in Figure 12.

A closer look at Figure 12 demonstrates that at the upstream undisturbed location ($\hat{x} = -4$), M_{30} and M_{12} begin with an insignificantly small negative value in the near-bed region and then consistently increased (negative magnitude) with a rise in $\bar{z}$ for both runs which are typical of turbulent shear flows over rigid surfaces [22]. Although, in the wake region, M_{30} and M_{12} start with small positive values, M_{30} and M_{12} decrease with a rise in the wall-normal distance from the bed and change their signs near the top level of the cylinder for both the runs. With a further rise in $\bar{z}$, their negative magnitude increase sharply and attain their negative peaks at $\bar{z} \approx 0.4$. After that, negative magnitudes of M_{30}

and M_{12} decrease gradually with $\overline{z}$ and become almost invariant of $\overline{z}$ in the free surface zone. It is noteworthy that the variation of M_{03} and M_{21} occur very similarly to M_{30} and M_{12}, however with the opposite signs. Unlike M_{30} and M_{12} distributions, M_{03} and M_{21} begin at bed with small negative values in the wake region. The negative magnitudes of M_{03} and M_{21} decrease with a further rise in $\overline{z}$, and change their signs at $z \approx$ D. After that, the value of M_{03} and M_{21} increase gradually with a rise in the vertical distance from the bed and attain peaks and decrease with a further rise in z, and become almost constant and invariant of z in the near free surface region. The positive and negative values of M_{30} and M_{03}, respectively, reveal that the downstream flux of streamwise RNS was associated with a downward flux of wall-normal RNS in the near-bed region, while the positive and negative values of M_{12} and M_{21} indicate the downstream advection of streamwise RNS and downward advection of wall-normal RNS in the near-bed region. Therefore, it can be understood that the dominance of sweep events in the near-bed region makes M_{30} and M_{12} to be positive and M_{03} and M_{21} to be negative. The negative values of M_{30} and M_{12} and positive values of M_{03} and M_{21} away from the boundary indicates the influence of the ejection events in the disturbed outer layer region. The sweep event represents a high-speed fluid parcel rushed towards the bed and is mainly responsible for bed mobility. From overall observations, it is clear that the enhanced contribution of M_{30} (positive) and M_{03} (negative) in the downstream region of the pipe reveal an intense momentum exchange in the near-bed region.

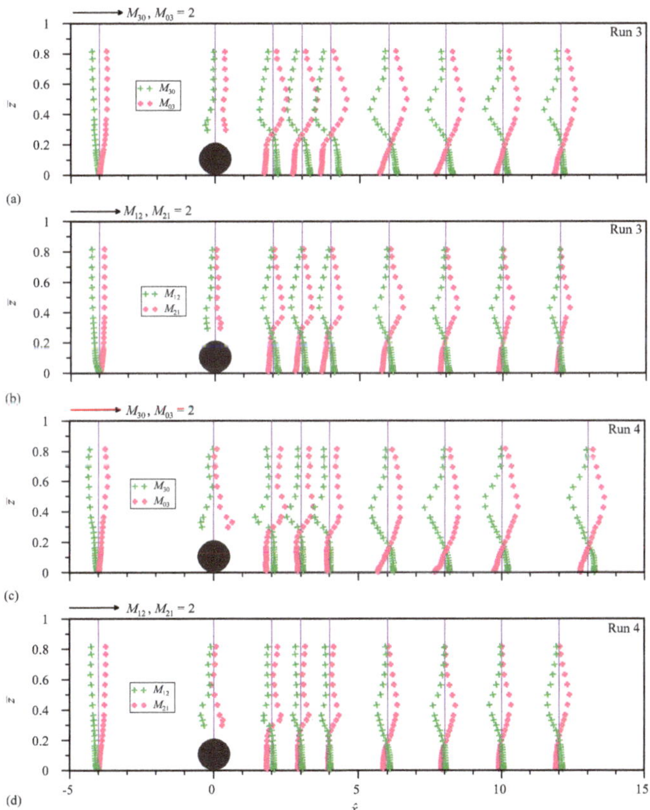

Figure 12. The wall-normal profiles of the third-order correlations, M_{ij} in the wake region of the pipe at different $\hat{x}$ for Run 3: (**a**) M_{30} and M_{03}, (**b**) M_{21} and M_{12}; and for Run 4: (**c**) M_{30} and M_{03}, and (**d**) M_{21} and M_{12}.

4. Conclusions

The significance of this research is that it analyze the self-preserving characteristics of turbulent flow statistics in the wall-wake of the bed-mounted horizontal pipe at different hydraulic and physical conditions. Additionally, the streamwise variations of the wall-normal profiles of the third-order correlations were evaluated. The self-preserving characteristics of the wall-wake region of the pipe were evaluated based on the collapse of the non-dimensional wall-normal profiles of velocity deficit $\Delta u/(\Delta u)_{max}$, non-dimensional RSS deficit $\Delta\tau_{uw}/(\Delta\tau_{uw})_{max}$, non-dimensional streamwise turbulence intensity deficit $\Delta\sigma_u/(\Delta\sigma_u)_{max}$, and non-dimensional wall-normal turbulence intensity deficit $\Delta\sigma_w/(\Delta\sigma_w)_{max}$ at various streamwise distances, $\hat{x}$. The velocity deficit, RSS deficit, and turbulence intensities deficit distributions exhibited good collapse under this scaling, demonstrating the self-preserving behaviour of the mean and turbulent flow field. The variation of the half velocity profile width, z_1, and half RSS profile width, z_2, with streamwise distance indicated the expansion or contraction of the wake region in the streamwise direction. Therefore, it is clear that the wall-wake flow of a bed-mounted horizontal circular pipe shows the self-preserving behaviour under different inflow conditions. The findings of this study will be useful for the validation and improvement of numerical models, and quick prediction of the wall-wake flows for evaluation of self-preservation. Finally, the important findings of the study are enumerated as follows.

1. The recirculation region is stretched up to $\hat{x} = 7.4$, which is more than the value found in the existing literature on the topic.
2. The self-preserving characteristics such as turbulence intensities and Reynolds shear stresses start from $\hat{x} = 3$ and the flow field is found to be not yet recovered even at $\hat{x} = 12$.
3. In the wall normal direction, self-preserving of velocity profiles, RSS profiles, and turbulence intensities is found for $z/D < 2$ and self preservation is not found for $z/D > 2$.
4. The points of inflection ($d^2\overline{u}/dz^2 = 0$) in the individual u^+ profiles are observed near the top level of the pipe, $z/D \approx 1$. Therefore, RSS and turbulence intensities are attaining their peaks at $z/D \approx 1$.
5. The velocity deficit profile exhibits a single peak in its distribution at $z/z_1 \approx 0.25$, the RSS deficit profiles display a single peak at $z/z_1 \approx 0.6$, and the turbulence intensities deficit profiles display a single peak at $z/z_1 \approx 0.8$.
6. The streamwise decay of peak values of non-dimensional velocity defect, non-dimensional RSS defect, and non-dimensional turbulence intensity defect indicates recovery of flow.
7. The half RSS defect profile width, half velocity defect profile width, and the half turbulence defect profile width grow in the streamwise direction, which indicates weaning of the wake strength.
8. In the wake region, third-order moments M_{03} and M_{21} begin at the bed with small negative values, whereas M_{30} and M_{12} begin at the bed with small positive values. They change their signs at $z \approx$ D, which proves that sweep events are dominant below the level of the cylinder and ejection events are dominant above the level of the cylinder.

Author Contributions: K.D.: writing—original draft preparation, writing—review and editing, and data curation; P.R.H.: writing—original draft preparation, writing—review and editing, funding acquisition, project administration, data curation, and supervision; R.B.: writing—review and editing, data curation; J.H.P.: writing—review and editing, and data curation. All authors contributed to the work. All authors have read and agreed to the published version of the manuscript.

Funding: This research received no external funding.

Institutional Review Board Statement: Not applicable.

Informed Consent Statement: Not applicable.

Data Availability Statement: The data presented in this study are available on reasonable request from the corresponding author.

Conflicts of Interest: The authors declare no conflict of interest.

References

1. Djeridi, H.; Braza, M.; Perrin, R.; Harran, G.; Cid, E.; Cazin, S. Near-wake turbulence properties around a circular cylinder at high Reynolds number. *Flow Turbul. Combust.* **2003**, *71*, 19–34. [CrossRef]
2. Konstantinidis, E.; Balabani, S.; Yianneskis, M. The effect of flow perturbations on the near wake characteristics of a circular cylinder. *J. Fluids Struct.* **2003**, *18*, 367–386. [CrossRef]
3. Braza, M.; Perrin, R.; Hoarau, Y. Turbulence properties in the cylinder wake at high Reynolds numbers. *J. Fluids Struct.* **2006**, *22*, 757–771. [CrossRef]
4. Akoz, M.S. Flow structures downstream of the horizontal cylinder laid on a plane surface. *Proc. Inst. Mech. Eng. Part C J. Mech. Eng. Sci.* **2009**, *223*, 397–413. [CrossRef]
5. Akoz, M.S.; Kirkgoz, M.S. Numerical and experimental analyses of the flow around a horizontal wall-mounted circular cylinder. *Trans. Can. Soc. Mech. Eng.* **2009**, *33*, 189–215. [CrossRef]
6. Devi, K.; Hanmaiahgari, P.R. Experimental analysis of turbulent open channel flow in the near-wake region of a surface-mounted horizontal circular cylinder. In *River Flow 2020: Proceedings of the 10th Conference on Fluvial Hydraulics, Delft, The Netherlands, 7–10 July 2020*; CRC: Boca Raton, FL, USA, 2020; pp. 194–202.
7. Devi, K.; Hanmaiahgari, P.R.; Balachandar, R.; Pu, J.H. A Comparative Study between Sand- and Gravel-Bed Open Channel Flows in the Wake Region of a Bed-Mounted Horizontal Cylinder. *Fluids* **2021**, *6*, 239. [CrossRef]
8. Balachandar, R.; Ramachandran, S.; Tachie, M.F. Characteristics of shallow turbulent near wakes at low Reynolds numbers. *J. Fluids Eng. Trans. ASME Eng.* **2000**, *122*, 302–308. [CrossRef]
9. Balachandar, R.; Tachie, M.F.; Chu, V.H. Concentration profiles in shallow turbulent wakes. *J. Fluids Eng. Trans. ASME* **1999**, *121*, 34–43. [CrossRef]
10. Nasif, G.; Barron, R.M.; Balachandar, R. DES evaluation of near-wake characteristics in a shallow flow. *J. Fluids Struct.* **2014**, *45*, 153–163. [CrossRef]
11. Akilli, H.; Rockwell, D. Vortex formation from a cylinder in shallow water. *Phys. Fluids* **2002**, *14*, 2957–2967. [CrossRef]
12. Singha, A.; Balachandar, R. Structure of wake of a sharp-edged bluff body in a shallow channel flow. *J. Fluids Struct.* **2011**, *27*, 233–249. [CrossRef]
13. Townsend, A. Measurements in the turbulent wake of a cylinder. *Proc. R. Soc. Lond. Ser. A Math. Phys. Sci.* **1947**, *190*, 551–561. [CrossRef]
14. Townsend, A. Momentum and energy diffusion in the turbulent wake of a cylinder. *Proc. R. Soc. Lond. Ser. A Math. Phys. Sci.* **1949**, *197*, 124–140. [CrossRef]
15. Maji, S.; Hanmaiahgari, P.R.; Balachandar, R.; Pu, J.H.; Ricardo, A.M.; Ferreira, R.M. A Review on Hydrodynamics of Free Surface Flows in Emergent Vegetated Channels. *Water* **2020**, *12*, 1218. [CrossRef]
16. Sadeque, M.F.; Rajaratnam, N.; Loewen, M.R. Shallow turbulent wakes behind bed-mounted cylinders in open channels. *J. Hydraul. Res.* **2009**, *47*, 727–743. [CrossRef]
17. Tachie, M.F.; Balachandar, R. Shallow wakes generated on smooth and rough surfaces. *Exp. Fluids* **2001**, *30*, 467–474. [CrossRef]
18. Maji, S.; Pal, D.; Hanmaiahgari, P.R.; Gupta, U.P. Hydrodynamics and turbulence in emergent and sparsely vegetated open channel flow. *Environ. Fluid Mech.* **2017**, *17*, 853–877. [CrossRef]
19. Pu, J.H. Velocity Profile and Turbulence Structure Measurement Corrections for Sediment Transport Induced Water-Worked Bed. *Fluids* **2021**, *6*, 86. [CrossRef]
20. Goring, D.G.; Nikora, V.I. Despiking Acoustic Doppler Velocimeter Data. *J. Hydraul. Eng.* **2002**, *128*, 117–126. [CrossRef]
21. Wahl, T.L. Discussion of "Despiking Acoustic Doppler Velocimeter Data" by Derek G. Goring and Vladimir I. Nikora. *J. Hydraul. Eng.* **2003**, *129*, 484–487. [CrossRef]
22. Mori, N.; Suzuki, T.; Kakuno, S. Noise of Acoustic Doppler Velocimeter Data in Bubbly Flows. *J. Eng. Mech.* **2007**, *133*, 122–125. [CrossRef]
23. Chanson, H.; Trevethan, M.; Aoki, S. Acoustic Doppler velocimetry (ADV) in small estuary: Field experience and signal post-processing. *Flow Meas. Instrum.* **2008**, *19*, 307–313. [CrossRef]
24. Nezu, I.; Nakagawa, H. *Turbulence in Open Channel Flows*; IAHR: Rotterdam, The Netherlands, 1993; ISBN 9054101180.
25. Gad-el-Hak, M.; Bandyopadhyay, P.R. Reynolds Number Effects in Wall-Bounded Turbulent Flows. *Appl. Mech. Rev.* **1994**, *47*, 307–365. [CrossRef]

MDPI

Article

A Comparative Study between Sand- and Gravel-Bed Open Channel Flows in the Wake Region of a Bed-Mounted Horizontal Cylinder

Kalpana Devi [1], Prashanth Reddy Hanmaiahgari [1,*], Ram Balachandar [2] and Jaan H. Pu [3,*]

[1] Department of Civil Engineering, IIT Kharagpur, Kharagpur 721302, India; kalpanarajpoot@iitkgp.ac.in
[2] Department of Civil and Environmental Engineering, University of Windsor, Windsor, ON N9B 3P4, Canada; rambala@uwindsor.ca
[3] School of Engineering, Faculty of Engineering and Informatics, University of Bradford, Bradford BD71DP, UK
* Correspondence: hpr@civil.iitkgp.ac.in (P.R.H.); j.h.pu1@bradford.ac.uk (J.H.P.)

Abstract: In nature, environmental and geophysical flows frequently encounter submerged cylindrical bodies on a rough bed. The flows around the cylindrical bodies on the rough bed are very complicated as the flow field in these cases will be a function of bed roughness apart from the diameter of the cylinder and the flow velocity. In addition, the sand-bed roughness has different effects on the flow compared to the gravel-bed roughness due to differences in the roughness heights. Therefore, the main objective of this article is to compare the mean velocities and turbulent flow properties in the wake region of a horizontal bed-mounted cylinder over the sand-bed with that over the gravel-bed. Three experimental runs, two for the sand-bed and one for the gravel-bed with similar physical and hydraulic conditions, were recorded to fulfil this purpose. The Acoustic Doppler Velocimetry (ADV) probe was used for measuring the three-dimensional (3D) instantaneous velocity data. This comparative study shows that the magnitude of mean streamwise flow velocity, streamwise Reynolds normal stress, and Reynolds shear stress are reduced on the gravel-bed compared to the sand-bed. Conversely, the vertical velocities and vertical Reynolds normal stress are higher on the gravel-bed than the sand-bed.

Keywords: ADV; bed-mounted horizontal cylinder; gravel-bed; sand-bed; turbulence; wake region

Citation: Devi, K.; Hanmaiahgari, P.R.; Balachandar, R.; Pu, J.H. A Comparative Study between Sand- and Gravel-Bed Open Channel Flows in the Wake Region of a Bed-Mounted Horizontal Cylinder. *Fluids* **2021**, *6*, 239. https://doi.org/10.3390/fluids6070239

Academic Editor: Mehrdad Massoudi

Received: 14 May 2021
Accepted: 26 June 2021
Published: 1 July 2021

Publisher's Note: MDPI stays neutral with regard to jurisdictional claims in published maps and institutional affiliations.

1. Introduction

Turbulent flows over rough surfaces occur in engineering applications and natural sciences. Determining the roughness effects on the turbulent flows in engineering applications is very important since the mechanisms of production, diffusion, and energy transfer between the mean and the turbulent fields, especially in the near-wall region, are influenced by surface roughness.

There has been considerable work carried out to understand the dynamics of turbulent flows over rough surfaces. The numerous studies in the past, before 1990, are related to the universal aspects of rough-walled flows. Most of them have highlighted the effects of uniformly distributed roughness elements of standard shapes, such as spheres, bars, racks, cylinders, ribs, dunes, and vegetation [1–6]. Though, very recently, a few researchers have focused on illustrating the realistic roughness effects on various features of turbulent flows [7,8]. The realistic roughness is quite different from the regular (or 'modelled') roughness. It is described by a wider spectrum of wavelengths and random distribution of structures of each scale.

Robert et al. [8] and Bigillon et al. [9] analyzed the turbulence statistics on the transitionally rough bed, while Bergstrom et al. [10] investigated the effects of surface roughness on the mean velocity distribution in a turbulent boundary layer. Bergstrom et al. [10] found that the outer flow was significantly affected by the roughness properties. Volino et al. [11]

used two-dimensional roughness fabricated by laying the transverse bars on the channel bed to determine the turbulence structure in a boundary layer. Bomminayuni and Stoesser [12] used a channel bed artificially roughened by hemispheres to analyze the turbulent statistics in the flow. The roughness effects on the near-wall turbulence in terms of Reynolds stress budget were determined numerically by Yuan and Piomelli [13]. Essel and Tachie [14] examined the wall roughness effects on the turbulent flow field downstream of the backward-facing step. They observed a significant reduction in the longitudinal and vertical spatial coherence of the turbulence structures due to the surface roughness. Wu and Piomelli [15] investigated the effects of surface roughness on the separating boundary layer over a flat plate by using large-eddy simulations.

Flow around submerged cylindrical bodies occurs in many environmental and geophysical systems—for example, water, oil, and gas pipelines, a communication cable laid on a river or sea bed, fish habitats, submerged horizontal or vertical pipes, sewer pipelines, hydraulic structures, etc. The flow field around a cylindrical object on a surface has been experimentally and numerically studied extensively in the past for many years. Though, most of the studies determined the wall-proximity effects on the flow characteristics such as the point of separation of wall-boundary layer [16], the drag and lift coefficients [17], the location of front stagnation point [18], distribution of the pressure on the cylinder surface [19], vortical flow structures [20], Strouhal number and vortex shedding phenomenon and flow [21], and turbulence properties of the wake region of the cylinder [22]. A few investigations have analyzed the turbulent flow field of a bed-mounted horizontal cylinder [23–25].

When the cylinder is laid on the rough bed like the sand-bed and gravel-bed, the complexities of the flow may enhance many folds since the flow, in this case, will be a function of bed roughness apart from the diameter of the cylinder and the flow velocity. In addition to that, the flow field on the sand-bed might be quite different from the gravel-bed due to differences in properties of bed roughness.

Due to their practical importance and complexity, several experimental investigations have been carried out to enhance our knowledge of the rough wall turbulent flows. However, there are very few studies that show the effects of gravel-bed roughness on the flow field of a circular cylinder. As per the author's knowledge, there is no comparative investigation between sand and gravel-bed to date to show the effects of bed roughness heterogeneity on the wake flow of the bed-mounted horizontal cylinder. From the literature survey, it is clear that no experimental data are available to understand the bed roughness heterogeneity effects on the mean velocities and turbulent quantities of flow over the cylinder. In addition to that, experimental data are needed to validate the numerical models. Therefore, the present research aims to analyze the roughness heterogeneity effects on the wake region of a bed-mounted circular cylinder in terms of the mean velocities and turbulence quantities.

2. Experimental Details

2.1. Experimental Instrumentations

Experiments were performed in a recirculating rectangular flume (narrow open channel having an aspect ratio less than 5) with dimensions of $L \times B \times H = 12 \times 0.91 \times 0.70$; $12 \times 0.91 \times 0.70$ [25] and $10 \times 0.60 \times 0.65$ for the sand-bed (Run 1 and Run 2) and gravel-bed (GB 1), respectively. The channel bed was provided with a constant longitudinal slope (S_0) of 0.23%, 0.23%, and 0.22% throughout the flume length for Run 1, Run 2, and GB 1, respectively. The experimental set up for Run 1 and Run 2 is the same as used by Devi and Hanmaiahgari [25]. The schematic diagram of the experimental setup with a coordinate system for GB 1 is shown in Figure 1. Photographs of flumes for sand-bed and gravel-bed experiments are shown in Figures 2 and 3. The flume has transparent glass sidewalls to facilitate the visual observation of the flow. The inflow discharge was controlled by a valve and metered by a calibrated V-notch weir located at the inlet tank, upstream of the stilling basin through which the flow enters the flume. The flow depth in the flume was adjusted by controlling tailgate water depth using a tailgate located at the downstream end of the flume. The cylinder was laid on the flume, covering the entire width of the flume at a distance of 6.5, 6.5, and 5.82 m from the flume inlet for Run 1, Run 2, and GB 1, respectively. The test section was chosen at a distance of 6.3, 6.3, and 5.5 m downstream of the inlet for Run 1, Run 2, and GB 1, respectively. It was ensured that the flow was fully developed in the test section.

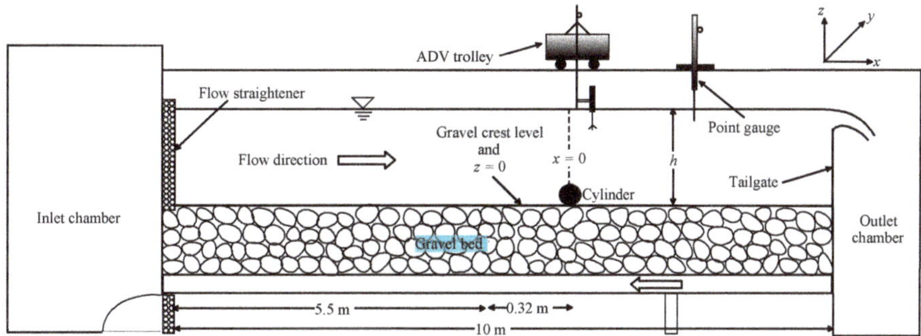

Figure 1. Schematic diagram of the experimental setup for the gravel-bed (GB 1).

Figure 2. The flume setup for the sand-bed (Run 1 and Run 2).

Figure 3. The flume setup for the gravel-bed (GB 1).

In this study, the x, y, and z axes were oriented in the longitudinal, spanwise, and vertical directions, respectively. The origin of the coordinate system was located at the bottommost point of the cylinder at the central vertical plane in the case of the sand-bed. The origin for the z-axis is the crest level of the top layer of the gravel particles. The nadir of the cylinder is in contact with the crest level. The z is taken positive vertically upwards and negative in the vertically downward direction from the crest level. The bed elevation was measured along the centerline of the flume. The coordinate system for GB 1 along with the schematic diagram of the experimental set up is shown in Figure 1. The time-averaged components of velocity are denoted by $\overline{u}$, $\overline{v}$, and $\overline{w}$ while u, v, and w represent the instantaneous velocity components in the x, y, and z directions, respectively. Here, u', v', and w' symbolize the fluctuating velocity components in the respective directions.

2.2. Bed Settings

The sand-bed (Run 1 and Run 2) was prepared by coating the uniform sand (d_{50} = 2.54 mm) over the flume bed. The gravel-bed (GB 1) was fabricated by arbitrarily spreading the well-sorted gravel of median size, d_{50} = 42.00 mm in four layers over the flume bed. The gravel particles were spread over the flume bed layer by layer in a compact and packed form. All the gravel particles crests over the gravel-bed were not at the same level as can be seen from the flume set up depicted in Figure 3. The geometric mean size, d_g $(= (d_{84.1}d_{15.9})^{0.5})$ for the sand and gravel are 2.5 and 42.7 mm, respectively. The geometric standard deviation σ_g $(= \sqrt{d_{84.1}/d_{15.9}})$ for both the sand and gravel material was calculated as 1.1. The gradation coefficient G $(= \frac{1}{2}(d_{84.1}/d_{50} + d_{50}/d_{15.9}))$ for the sand sample was 1.10 (< 1.4), which implies that it is uniform sand. Here, $d_{84.1}$ and $d_{15.9}$ represent the size of sediment for which 84.1% and 15.9%, respectively, of the mixture are finer. The particle size distribution curves for the sand-bed (Run 1 and Run 2) and gravel-bed are shown in Figure 4a,b, respectively. A closer look at Figure 4 shows that the sediment size in the case of the gravel varies over a wider range as compared to the sand. The median size (d_{50}) for the sand and gravel samples are demarcated in Figure 4.

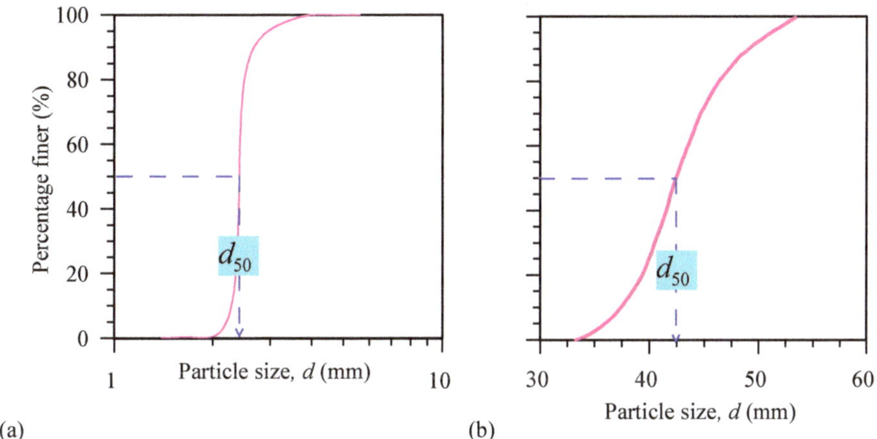

Figure 4. Particle size distribution curve for (**a**) the sand-bed (Run 1 and Run 2) and (**b**) the gravel-bed material (GB 1).

2.3. Experimental Conditions and Measuring Stations

The experiments were conducted under a uniform steady flow condition in a narrow open channel (aspect ratio $Ar < 5$). The experiments were conducted for flow depths (h) of 0.30, 0.30, and 0.25 m for Run 1, Run 2, and GB 1, respectively (Table 1). The flow depth was measured with the help of a point gauge with an accuracy of $\pm$ 0.1 mm by a Vernier scale attachment. The approach flow shear velocity (u_*), which estimates the bed resistance to the flow, was obtained by extrapolating the RSS profile to the flume bed [26]. The values of approach flow shear velocity (u_*) at upstream location ($x = -4D$) for Run 1, Run 2, and GB 1 are 7.1, 9.2, and 27 mm/s, respectively. The hydraulic and physical parameters of all experimental runs (Run 1, Run 2, and GB 1) are given in Table 1. Here, the aspect ratio (Ar) is defined as the ratio of channel width (B) to flow depth (h). For GB 1, the flow depth h was measured with respect to the crest level of the gravel particles, z = 0. Fr is the Froude number ($= \frac{U}{\sqrt{gh}}$) and Re is the flow Reynolds number ($Re = Uh/v$). The shear Reynolds number (R_*) is calculated using Equation (1):

$$R_* = k_s u_* / v \tag{1}$$

where k_s is the equivalent roughness height, and v is the kinematics viscosity of water. Here the median particle size, d_{50} was considered equivalent to the equivalent roughness height, k_s.

Table 1. Hydraulic and physical parameters of all experimental runs (Run 1, Run 2, and GB 1).

Exp. Run	D (m)	S_0(%)	h (m)	U (m/s)	Ar	d_{50}(m)	Re	Re_*	Fr
Run 1	0.05	0.023	0.30	0.15	3.1	0.00254	45,000	18.03	0.09
Run 2	0.05	0.023	0.30	0.19	3.1	0.00254	57,000	23.37	0.11
GB 1	0.06	0.022	0.25	0.25	2.4	0.04200	62,500	1050	0.16

Velocity profiles were measured in the central vertical plane (xz-plane) of the channel at nine ($x = -4D$, 0, $2D$, $3D$, $4D$, $6D$, $8D$, $10D$, and $12D$) different measuring stations in the longitudinal direction of the flow for all runs. At $x = -4D$, the flow is undisturbed, and it is located upstream of the cylinder. The cylinder is fixed at streamwise location $x = 0$.

2.4. ADV System and Data Collection Procedure

The instantaneous flow velocities were measured by a Three-Dimensional Acoustic Doppler Velocimetry (3D ADV) instrument. A four-receiver down-looking ADV probe, named Vectrino plus (manufactured by Nortek), was used, which significantly reduces the noise signal of the measurements compared to a three-receivers probe [27]. It was functioning with an acoustic frequency of 10 MHz to capture the instantaneous 3D flow velocities, and the sampling rate was 100 Hz as used by Maji et al. [28]. Since the measuring location was 5 cm below the probe, the data acquisition was not possible near the free water surface.

It was ensured that the sampling volume did not touch the flume bed during data collection. The sampling duration was taken as 300 s to ensure that the time-averaged velocities are statistically time-independent. During the experiments, the least value of the signal-to-noise (SNR) ratio and the correlation coefficient were retained as 17 and 70, respectively. The signal captured by the Vectrino in the near-bed flow zone contained spikes due to the interaction between the incident and reflected pulses. Therefore, the raw data were filtered by a spike removal algorithm, known as the phase-space threshold method developed by Goring and Nikora [29].

3. Results and Discussion

The time-averaged longitudinal ($\overline{u}$) and vertical flow velocities ($\overline{w}$) were estimated as an ensemble average of the instantaneous velocities of the respective directions and mathematically expressed by Equations (2) and (3) as used by Devi and Hanmaiahgari [25]:

$$\overline{u} = \frac{1}{N}\sum_{i=1}^{N} u_i \tag{2}$$

$$\overline{w} = \frac{1}{N}\sum_{i=1}^{N} w_i \tag{3}$$

where u_i and w_i indicate the longitudinal and vertical components of instantaneous velocity, respectively, and N is the total number of data samples collected. The method of Reynolds decomposition gives the components of fluctuating velocity in the longitudinal (u') and vertical directions (w') by Equation (4) and (5), respectively as:

$$u' = u_i - \overline{u} \tag{4}$$

$$w' = w_i - \overline{w} \tag{5}$$

Reynolds normal stresses in the longitudinal (σ_{uu}) and vertical (σ_{ww}) directions are expressed as the mean of quadratic product of the fluctuating components of the velocities in the corresponding directions and are evaluated by Equations (6) and (7):

$$\sigma_{uu} = \overline{u'u'} = \frac{1}{N}\sum_{i=1}^{N}(u_i - \overline{u})^2 \tag{6}$$

$$\sigma_{ww} = \overline{w'w'} = \frac{1}{N}\sum_{i=1}^{N}(w_i - \overline{w})^2 \tag{7}$$

Reynolds shear stress per unit mass (τ_{uw}) was estimated by using Equation (8).

$$\tau_{uw} = -\overline{u'w'} = -\frac{1}{N}\sum_{i=1}^{N}(u_i - \overline{u})(w_i - \overline{w}) \tag{8}$$

The maximum longitudinal time-averaged velocity occurs within a 5 cm depth from the water surface. Due to the limitations of the ADV probe, it was not possible to collect data within a 5 cm depth from the water surface. It is assumed that the free surface effect in the subcritical flows is limited to a very shallow region near the free surface and it will not change underlying hydrodynamics. Therefore, the approach flow maximum velocity (U_{max}) is taken as the maximum velocity of each run at $x = -4D$ from the available data measuring points. The mean and turbulent flow characteristics for all three runs were normalized by the approach flow maximum velocity (U_{max}) and the vertical distances (z) were normalized by the flow depth (h). The normalized vertical distance is denoted by $\overline{z}$ $(= z/h)$. The normalized mean velocities, normalized RNS, and normalized RSS were plotted against normalized vertical distance ($\overline{z}$) for Run 1 and Run 2 together, and GB 1 separately at all nine measuring stations. The plots are analyzed and compared to examine the surface roughness effects on the wake region of the cylinder.

The entire flow field of the cylinder is divided into three flow zones, namely undisturbed upstream, recirculation region, and redevelopment region, as used by Essel and Tachie [30] in the case of forward-facing step. The recirculation and redevelopment regions are in the wake region of the cylinder. The measuring station $x = -4D$ is located upstream of the cylinder; $x = 2D, 3D, 4D$, and $6D$ are in the recirculation region; and $x = 8D, 10D$, and $12D$ fall in the redevelopment region. It is noteworthy to mention that the flow is still under the recovery process at $x = 12D$, and it is not yet fully recovered. A schematic of the upstream, recirculation, and redevelopment regions of a circular cylinder is shown in Figure 5.

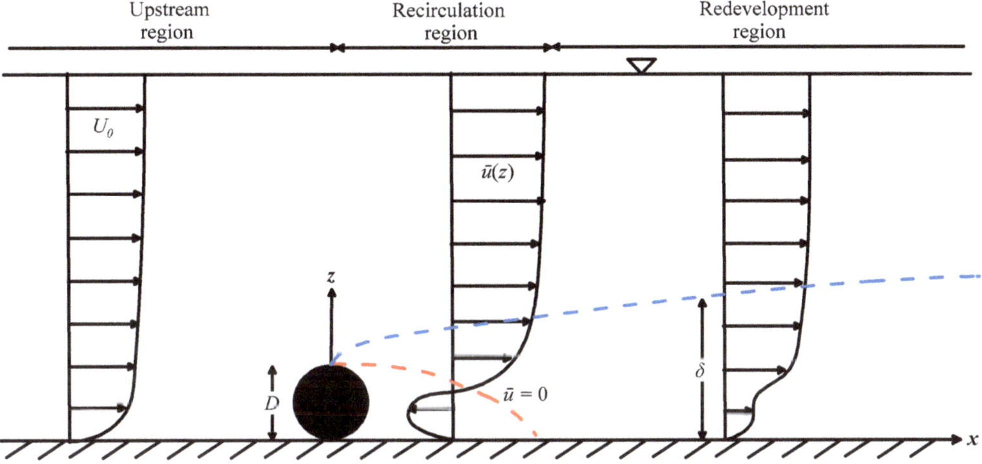

Figure 5. Schematic of different regions of flow field upstream and downstream of a circular cylinder.

The main findings of the proposed experimental research are discussed below.

3.1. Mean Velocities

The time-averaged velocities in the longitudinal ($\overline{u}$) and vertical directions ($\overline{w}$) are normalized by the approach flow maximum velocity (U_{max}) and are written as $\overline{u}/U_{max}$ and $\overline{w}/U_{max}$, respectively. Figure 6 examines the vertical profiles of normalized time-averaged longitudinal velocity ($\overline{u}/U_{max}$) for Run 1, Run 2, and GB 1.

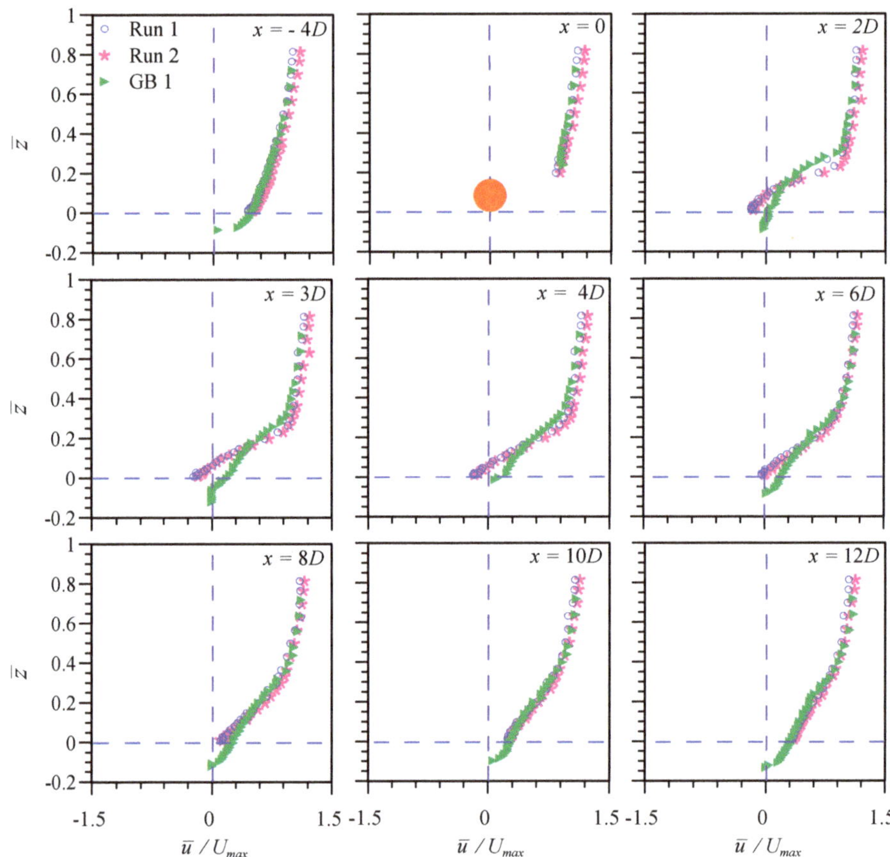

Figure 6. The vertical profiles of normalized time-averaged longitudinal velocity ($\overline{u}/U_{max}$) at different measuring stations (x) for Run 1, Run 2, and GB 1.

From Figure 6, it is clearly visible that the streamwise velocity is higher for Run 2 compared to Run 1. The former case has a higher Reynolds number (Re) compared to the latter, resulting in higher streamwise velocity. It is also observed that at upstream undisturbed location ($x = -4D$), the vertical profile of $\overline{u}/U_{max}$ for both types of beds follows logarithmic law. All the velocity profiles in the wake region preserve no-slip conditions. No negative longitudinal velocity is observed in the near-bed flow of the recirculation region over the gravel-bed, which shows a diminished separation zone on the gravel-bed. This observation is contrary to the observation made in the case of the sand-bed, where the longitudinal velocity in the near-bed flow of the recirculation region is negative in magnitude. This comparative study shows that the magnitude of mean streamwise flow velocity is smaller on the gravel-bed than the sand-bed, although the Reynolds number is higher for the flow on the gravel-bed (Figure 6). Figure 6 infers that the gravel-bed profiles are 'less full' than the sand-bed profiles. This phenomenon is similar to a reduction in longitudinal velocity due to roughness upstream and downstream of the rough backward-facing step (Wu et al. [31]). A closer look at Figure 6 shows that the vertical location of a point of inflection ($d^2u/dz^2 = 0$) in the $\overline{u}/U_{max}$ distribution is higher in the case of the gravel-bed ($\overline{z} \approx 0.30$) as compared to the sand-bed ($\overline{z} \approx 0.25$). Hence, it is concluded that the vertical location of the point of inflection ($d^2u/dz^2 = 0$) for the gravel-bed shifted away from the bed, implying a decrease in the longitudinal near-bed velocity caused by the gravel-bed roughness and increase in the thickness of roughness sublayer on the gravel-bed.

Figure 7 depicts the variation of normalized time-averaged vertical velocity ($\overline{w}/U_{max}$) against normalized vertical distance ($\overline{z}$) at different measuring stations (x) for Run 1, Run 2, and GB 1. It is very clear from Figure 7 that the vertical velocity is higher for Run 2 compared to Run 1 throughout the vertical domain. In the upstream of the cylinder ($x = -4D$), on the sand-bed, $\overline{w}/U_{max}$ starts with a small positive value and increases very slowly with $\overline{z}$ and becomes almost constant with $\overline{z}$ above $\overline{z} \approx 0.2$. Unlike the sand-bed, in the case of gravel, the $\overline{w}/U_{max}$ starts below the gravel-bed with a small positive value, then suddenly changes to a negative value and then with a further rise in $\overline{z}$, its negative value declines, and it becomes positive for $\overline{z} \geq 0.06$. Thereafter, the vertical velocity increases slowly with vertical distance and become almost invariant of $\overline{z}$ like over the sand-bed. It is also observed that for $x = -4D$, 0, and $2D$, the profiles are positive on both the sand and gravel-beds with an exception at $x = -4D$ on the gravel-bed. As already mentioned, at $x = -4D$ on the gravel-bed, below the bed level velocity is negative up to $\overline{z} \approx 0.06$. In the wake region, the vertical velocity starts with a small negative value and then increases (negative value) abruptly with a rise in $\overline{z}$ and attains a negative peak. Thereafter, the negative value quickly declines with an increase in vertical distance from the bed and become almost constant with $\overline{z}$ variation above $\overline{z} \approx 0.4$. These patterns in the $\overline{w}/U_{max}$ distribution is similar on the sand and gravel-beds for the wake region with an exception at $x = 2D$. At $x = 2D$, for both beds, the $\overline{w}/U_{max}$ profiles are completely positive, unlike at other locations, where they are negative. The vertical location of the peak at $x = 2D$ on the sand-bed corresponds to the top level of the cylinder ($z \approx 1$), while this is not true over the gravel-bed. The vertical location of the peak of $\overline{w}/U_{max}$ over the gravel-bed has shifted away from the bed due to the damping of the longitudinal velocity resulting from the heterogeneity of the gravel-bed. In the recirculation region, the vertical location of the peak of $\overline{w}/U_{max}$ shifted towards the bed with the increase in x. On the contrary, in the redevelopment region, for both the sand and gravel-beds, the vertical location of the peak moved away from the bed with an increase in x. In the recirculation region, the magnitude of the negative peak of $\overline{w}/U_{max}$ increases in the streamwise direction with x. Conversely, the declining patterns in the magnitude of the negative peak with an increase in x are observed in the redevelopment region for both the sand and gravel-beds. The location of peak ($\overline{z} > 0.4$) on the gravel-bed is higher than the location of the peak ($\overline{z} < 0.4$) on the sand-bed. It is clearly observed from Figure 7 that the maximum negative vertical mean velocity on the sand-bed and gravel-bed is about $0.1U_{max}$ and $0.05U_{max}$, respectively. Hence, an inference is made that gravel-bed roughness has dampened the value of the maximum negative vertical mean velocity.

3.2. Reynolds Normal Stress (RNS) Distribution

The Reynolds normal stresses (RNSs) define the strength of the turbulence and give an idea about the fluctuating velocity components. The RNSs are affected by factors such as velocity, roughness size, roughness orientation, etc. These are denoted as σ_{uu} and σ_{ww} in the longitudinal (x) and vertical (z) directions, respectively. They are normalized by the approach flow maximum velocity ${U_{max}}^2$ and are given as $\sigma_{uu}/{U_{max}}^2$ and $\sigma_{ww}/{U_{max}}^2$, respectively. The normalized streamwise ($\sigma_{uu}/{U_{max}}^2$) and vertical ($\sigma_{ww}/{U_{max}}^2$) RNSs are plotted against the normalized vertical distance ($\overline{z}$) for all the three runs (Run 1, and Run 2, and GB 1) in Figures 8 and 9, respectively, at different measuring stations (x).

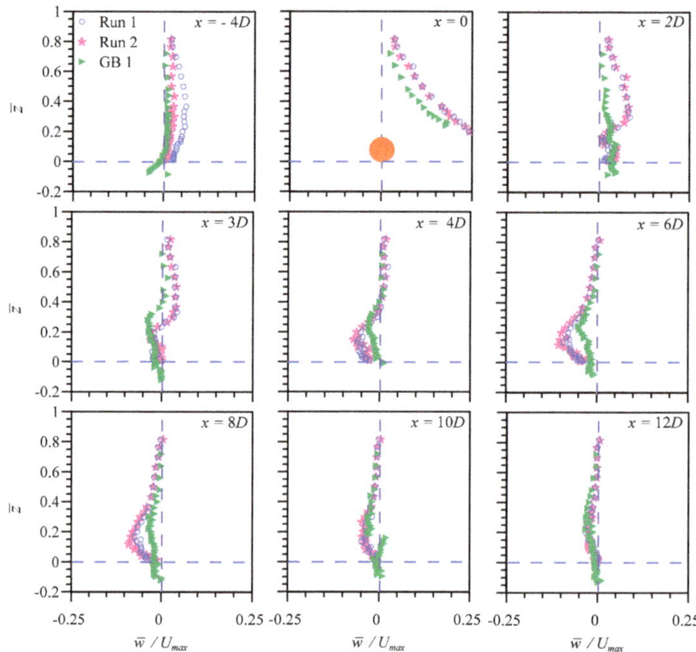

Figure 7. The variation of normalized time-averaged vertical velocity ($\overline{w}/U_{max}$) against normalized vertical distance ($\overline{z}$) at different measuring stations (x) for Run 1, Run 2, and GB 1.

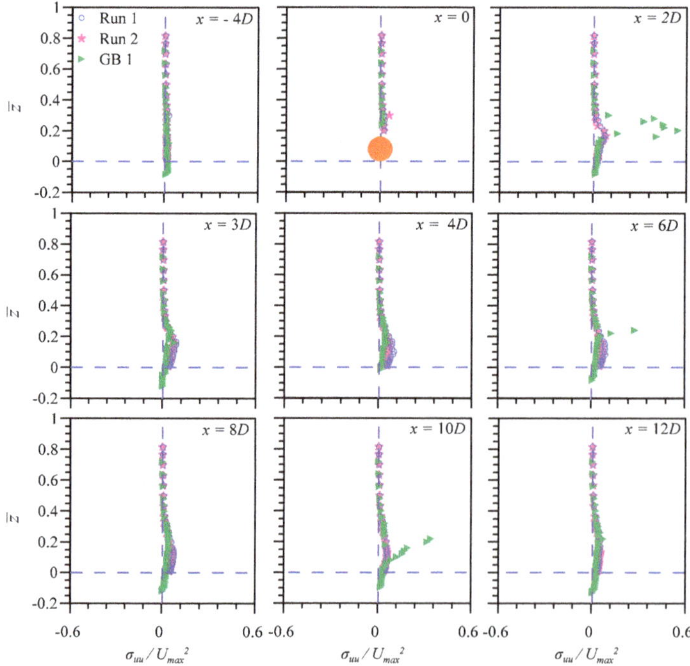

Figure 8. The vertical profiles of normalized longitudinal RNS ($\sigma_{uu}/U_{max}{}^2$) at different measuring stations (x) for Run 1, Run 2, and GB 1.

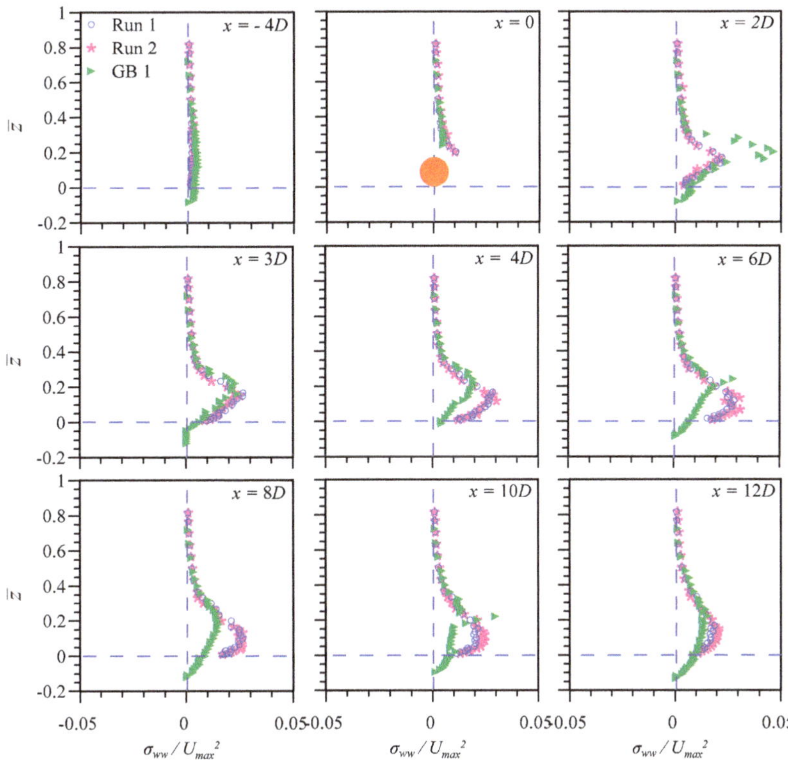

Figure 9. The variation of normalized time-averaged vertical RNS ($\sigma_{ww}/U_{max}{}^2$) against normalized vertical distance ($\overline{z}$) at different measuring stations (x) for Run 1, Run 2, and GB 1.

From Figures 8 and 9, it is noticed that the value of both σ_{uu} and σ_{ww} are higher in the case of Run 2 compared to Run 1. It is observed that in the uninterrupted upstream flow ($x = -4D$), the $\sigma_{uu}/U_{max}{}^2$ and $\sigma_{ww}/U_{max}{}^2$ profiles demonstrate diminishing trends against a rise in the $\overline{z}$ over the sand and gravel-beds. In the recirculation region, $\sigma_{uu}/U_{max}{}^2$ and $\sigma_{ww}/U_{max}{}^2$ start with a small positive value and increase sharply with the rise in $\overline{z}$ and attain a peak. Thereafter, it declines abruptly with further increase in $\overline{z}$ and then attains a small positive value and becomes almost invariant with $\overline{z}$. In the recirculation region ($x = 2D$, $3D$, $4D$, and $6D$), the vertical locations of the peaks of normal stresses are occurring near the bed. In contrast, they shifted away from the bed in the redevelopment region for both the sand- and gravel-bed flows. Their peaks were increasing in the recirculation region, but they decrease in the redevelopment region for both the sand- and gravel-bed flows. However, it is observed that the magnitudes of $\sigma_{uu}/U_{max}{}^2$ peaks on the gravel-bed are smaller than the sand-bed. Contrary to this, the peaks of $\sigma_{ww}/U_{max}{}^2$ on the gravel-bed are larger than the sand-bed. The gravel-bed roughness considerably diminished $\sigma_{uu}/U_{max}{}^2$ in both the recirculation and redevelopment regions compared to the sand-bed, which is similar to the observation of Wu et al. [31]. Although, $\sigma_{ww}/U_{max}{}^2$ are enhanced significantly on the gravel-bed as compared to the sand-bed in both the regions, which is contrary to the observation of Wu et al. [31]. In fact, Wu et al. [31] have observed that due to forward-facing step roughness, the RNS decreased downstream of the step. Furthermore, $\sigma_{uu}/U_{max}{}^2$ on the gravel-bed is more significantly diminished as compared to the increase in $\sigma_{ww}/U_{max}{}^2$ on the gravel-bed. In Figure 8, $\sigma_{uu}/U_{max}{}^2$ profiles over GB 1 show kinks at X/D = 2, 6, and 10 at z/h = 0.2, which is attributed to measurement error at that location. Finally, it is concluded that the peak normal stresses (turbulence intensities) show increasing trends in the recirculation region and decreasing patterns in the reattached region due to the enhanced turbulence in the former region and the attenuation of turbulence in the latter region.

3.3. Reynolds Shear Stress (RSS)

The Reynolds shear stress (RSS) is the time-averaged value of the quadratic terms of the fluctuating velocities components. The Reynolds shear stress has a similar influence on the flow as viscosity. The RSS offers resistance to the flow in the corresponding plane and governs secondary flows. The RSS in the xz plane per unit weight is denoted as τ_{uw} and the RSS normalized by the approach flow maximum velocity ($U_{max}{}^2$) is written as $\tau_{uw}/U_{max}{}^2$. The vertical profiles of normalized RSS ($\tau_{uw}/U_{max}{}^2$) at different measuring stations (x) for Run 1, Run 2, and GB 1 are shown in Figure 10.

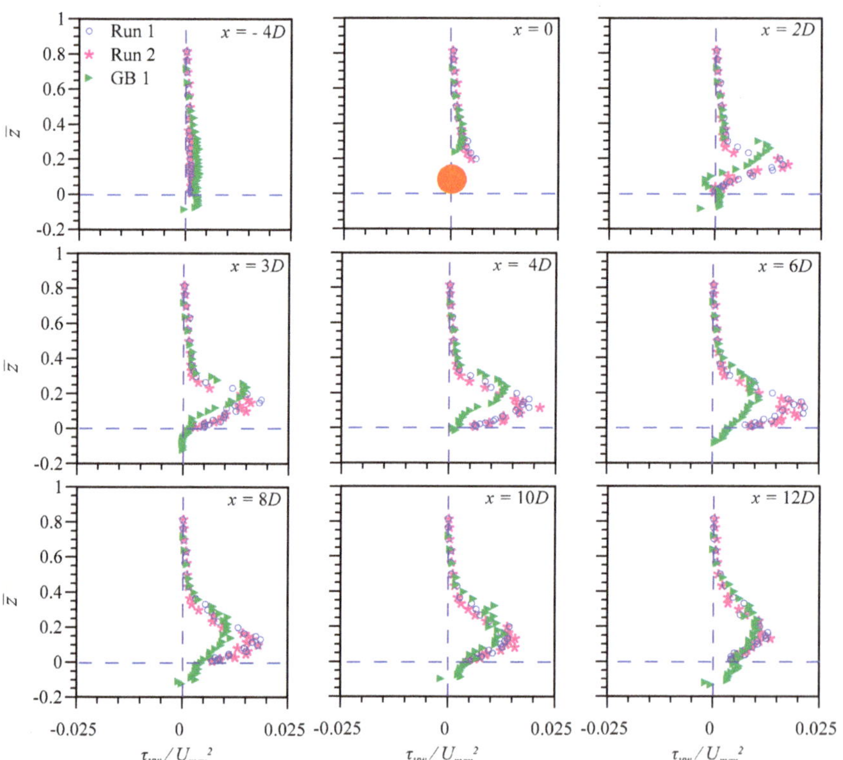

Figure 10. The vertical profiles of normalized RSS ($\tau_{uw}/U_{max}{}^2$) at different measuring stations (x) for Run 1, Run 2, and GB 1.

As evident from Figure 10, RSS is higher for Run 2 compared to Run 1 throughout the study domain. In the upstream of the cylinder, the RSS varies linearly with the vertical distance for both sand-bed and gravel-bed. The RSS magnitude in the wake region is slightly smaller on the gravel-bed than the sand-bed, although the Reynold number (*Re*) is considerably higher in the former case than the latter. The vertical location of the RSS peak is more elevated on the gravel-bed in the wake flow, which shows that roughness is extended farther in the gravel-bed flows. Interestingly, peaks of RSS are occurring at the top level of the cylinder on the sand-bed. In the recirculation region, $\tau_{uw}/U_{max}{}^2$ starts with a small positive value and increases sharply with a rise in the vertical distance and attains a peak much above the cylinder height. After that, it reduces sharply with a further rise in

$\bar{z}$ from the bed, then attains a minimal positive value and becomes almost invariant with $\bar{z}$. In the recirculation region ($x = 2D, 3D, 4D$, and $6D$), the vertical location of the peak of the RSS shifted towards the bed. In contrast, the peak is moved away from the bed for both the beds in the redevelopment region. The magnitude of the $\tau_{uw}/{U_{max}}^2$ peak increases in the recirculation region, but it decreases in the redevelopment region for both the sand- and gravel-bed flows. In the wake flow, the peak of RSS is occurring at $\bar{z} \approx 0.2$ for $x = 2D$ while moving away from the cylinder, and the vertical location of the peak is shifted towards the bed. The RSS peaks values are declining along the streamwise direction downstream of the cylinder over both the sand-bed and gravel-bed. This finding on the gravel-bed is similar to the flow pattern downstream of the rough forward-facing step (Wu et al. [31]).

4. Conclusions

This analysis conveys a clear understanding that gravel-bed roughness in comparison to sand-bed roughness significantly modifies the turbulence levels downstream of a horizontally mounted cylinder. In addition to that, the higher values of streamwise velocity, vertical velocity, RNSs, and RSS were observed in the case of a higher Reynolds number flow over the sand-bed due to a higher level of turbulence. In contrast, for similar Reynolds flow over sand- and gravel-beds, turbulence statistics such as RNSs and RSS are higher on the sand-bed than the gravel-bed. The peak values of streamwise velocity, vertical velocity, RSS, and RNSs have been dampened by the gravel-bed roughness. The enhanced roughness over the gravel-bed shifts the location of the peak of vertical velocity over the gravel-bed away from the flume bed compared to the location of the peak of vertical velocity over the sand-bed. Similarly, the vertical location of the point of inflection in the streamwise velocity profile is higher in the case of the gravel-bed than the sand-bed. The peak RNSs and RSS show increasing trends in the recirculation region and decreasing patterns in the reattached region over and sand- and gravel-beds. Finally, it is observed that the flow is mainly affected in the near-bed region downstream of a cylinder over and sand- and gravel-beds. The proposed research findings are important for the validation of numerical modelling results and improvement of the current numerical models.

Author Contributions: K.D.: writing—original draft preparation, writing—review and editing, and data curation; P.R.H.: writing—original draft preparation, writing—review and editing, funding acquisition, project administration, data curation, and supervision; R.B.: writing—review and editing, data curation, and funding acquisition; J.H.P.: writing—review and editing, funding acquisition and data curation. All authors contributed to the work. All authors have read and agreed to the published version of the manuscript.

Funding: The Author Ram Balachandar acknowledges the grant support from Natural Sciences and Engineering Research Council of Canada the author Jaan H. Pu acknowledges the grant support from the Hidden Histories of Environmental Science Project (at Seedgrant Stage) by the Natural Environment Research Council (NERC) and Arts and Humanities Research Council (AHRC), part of UK Research and Innovation (UKRI).

Institutional Review Board Statement: Not applicable.

Informed Consent Statement: Not applicable.

Data Availability Statement: The data presented in this study are available on reasonable request from the corresponding author.

Conflicts of Interest: The authors declare no conflict of interest.

References

1. Tachie, M.F.; Bergstrom, D.J.; Yang, Z.; Fang, X.; Wang, B.-C. Highly-disturbed turbulent flow in a square channel with V-shaped ribs on one wall. *Int. J. Heat Fluid Flow* **2015**, *56*, 182–197. [CrossRef]
2. Tsikata, J.M.; Tachie, M.F.; Katopodis, C. Open-channel turbulent flow through bar racks. *J. Hydraul. Res.* **2014**, *52*, 630–643. [CrossRef]
3. Balachandar, R.; Bhuiyan, F. Higher-Order Moments of Velocity Fluctuations in an Open-Channel Flow with Large Bottom Roughness. *J. Hydraul. Eng.* **2007**, *133*, 77–87. [CrossRef]

4. Djenidi, L.; Antonia, R.A.; Amielh, M.; Anselmet, F. A turbulent boundary layer over a two-dimensional rough wall. *Exp. Fluids* **2008**, *44*, 37–47. [CrossRef]
5. Krogstad, P.Å.; Andersson, H.I.; Bakken, O.M.; Ashrafian, A. An experimental and numerical study of channel flow with rough walls. *J. Fluid Mech.* **2005**, *530*, 327–352. [CrossRef]
6. Wu, Y.; Ren, H. On the impacts of coarse-scale models of realistic roughness on a forward-facing step turbulent flow. *Int. J. Heat Fluid Flow* **2013**, *40*, 15–31. [CrossRef]
7. Ren, H.; Wu, Y. Turbulent boundary layers over smooth and rough forward-facing steps. *Phys. Fluids* **2011**, *23*, 045102. [CrossRef]
8. Robert, A.; Roy, A.G.; Serres, B. De Turbulence at a roughness transition in a depth limited flow over a gravel bed. *Geomorphology* **1996**, *16*, 175–187. [CrossRef]
9. Bigillon, F.; Nino, Y.; Garcia, M. Measurements of turbulence characteristics in an open-channel flow over a transitionally-rough bed using particle image velocimetry. *Exp. Fluids* **2006**, *41*, 857–867. [CrossRef]
10. Bergstrom, D.J.; Kotey, N.A.; Tachie, M.F. The effects of surface roughness on the mean velocity profile in a turbulent boundary layer. *J. Fluids Eng. Trans. ASME* **2002**, *124*, 664–670. [CrossRef]
11. Volino, R.J.; Schultz, M.P.; Flack, K.A. Turbulence structure in a boundary layer with two-dimensional roughness. *J. Fluid Mech.* **2009**, *635*, 75–101. [CrossRef]
12. Bomminayuni, S.; Stoesser, T. Turbulence Statistics in an Open-Channel Flow over a Rough Bed. *J. Hydraul. Eng.* **2011**, *137*, 1347–1358. [CrossRef]
13. Yuan, J.; Piomelli, U. Roughness effects on the Reynolds stress budgets in near-wall turbulence. *J. Fluid Mech.* **2014**, *760*, 1–12. [CrossRef]
14. Essel, E.E.; Tachie, M.F. Roughness effects on turbulent flow downstream of a backward facing step. *Flow, Turbul. Combust.* **2015**, *94*, 125–153. [CrossRef]
15. Wu, W.; Piomelli, U. Effects of surface roughness on a separating turbulent boundary layer. *J. Fluid Mech.* **2018**, *841*, 552–580. [CrossRef]
16. Agbaglah, G.; Mavriplis, C. Three-dimensional wakes behind cylinders of square and circular cross-section: Early and long-time dynamics. *J. Fluid Mech.* **2019**, *870*, 419–432. [CrossRef]
17. Zhou, B.; Wang, X.; Guo, W.; Gho, W.M.; Tan, S.K. Experimental study on flow past a circular cylinder with rough surface. *Ocean Eng.* **2015**, *109*, 7–13. [CrossRef]
18. Cao, S.; Ozono, S.; Tamura, Y.; Ge, Y.; Kikugawa, H. Numerical simulation of Reynolds number effects on velocity shear flow around a circular cylinder. *J. Fluids Struct.* **2010**, *26*, 685–702. [CrossRef]
19. Cao, S.; Ozono, S.; Hirano, K.; Tamura, Y. Vortex shedding and aerodynamic forces on a circular cylinder in linear shear flow at subcritical Reynolds number. *J. Fluids Struct.* **2007**, *23*, 703–714. [CrossRef]
20. Akoz, M.S. Flow structures downstream of the horizontal cylinder laid on a plane surface. *Proc. Inst. Mech. Eng. Part C J. Mech. Eng. Sci.* **2009**, *223*, 397–413. [CrossRef]
21. Gu, F.; Wang, J.S.; Qiao, X.Q.; Huang, Z. Pressure distribution, fluctuating forces and vortex shedding behavior of circular cylinder with rotatable splitter plates. *J. Fluids Struct.* **2012**, *28*, 263–278. [CrossRef]
22. Ikegaya, N.; Morishige, S.; Matsukura, Y.; Onishi, N.; Hagishima, A. Experimental study on the interaction between turbulent boundary layer and wake behind various types of two-dimensional cylinders. *J. Wind Eng. Ind. Aerodyn.* **2020**, *204*, 104250. [CrossRef]
23. Akoz, M.S.; Sahin, B.; Akilli, H. Flow characteristic of the horizontal cylinder placed on the plane boundary. *Flow Meas. Instrum.* **2010**, *21*, 476–487. [CrossRef]
24. Kirkgoz, M.S.; Oner, A.A.; Akoz, M.S. Numerical modeling of interaction of a current with a circular cylinder near a rigid bed. *Adv. Eng. Softw.* **2009**, *40*, 1191–1199. [CrossRef]
25. Devi, K.; Hanmaiahgari, P.R. Experimental analysis of turbulent open channel flow in the near-wake region of a surface-mounted horizontal circular cylinder. In Proceedings of the River Flow 2020: Proceedings of the 10th Conference on Fluvial Hydraulics, Delft, The Netherlands, 7–10 July 2020; pp. 194–202.
26. Pu, J.H. Velocity Profile and Turbulence Structure Measurement Corrections for Sediment Transport-Induced Water-Worked Bed. *Fluids* **2021**, *6*, 86. [CrossRef]
27. Blanckaert, K.; Lemmin, U. Means of noise reduction in acoustic turbulence measurements. *J. Hydraul. Res.* **2006**, *44*, 3–17. [CrossRef]
28. Maji, S.; Pal, D.; Hanmaiahgari, P.R. Hydrodynamics and turbulence in emergent and sparsely vegetated open channel flow. *Environ. Fluid Mech.* **2017**, *17*, 853–877. [CrossRef]
29. Goring, D.G.; Nikora, V.I. Despiking Acoustic Doppler Velocimeter Data. *J. Hydraul. Eng.* **2002**, *128*, 117–126. [CrossRef]
30. Essel, E.E.; Tachie, M.F. Upstream roughness and Reynolds number effects on turbulent flow structure over forward facing step. *Int. J. Heat Fluid Flow* **2017**, *66*, 226–242. [CrossRef]
31. Wu, Y.; Ren, H.; Tang, H. Turbulent flow over a rough backward-facing step. *Int. J. Heat Fluid Flow* **2013**, *44*, 155–169. [CrossRef]

Article

Stability and Consolidation of Sediment Tailings Incorporating Unsaturated Soil Mechanics

Alfrendo Satyanaga [1], Martin Wijaya [2], Qian Zhai [3], Sung-Woo Moon [1], Jaan Pu [4] and Jong R. Kim [1,*]

1 Department of Civil and Environmental Engineering, Nazarbayev University, 53 Kabanbay Batyr Ave., Nur-Sultan 010000, Kazakhstan; alfrendo.satyanaga@nu.edu.kz (A.S.); sung.moon@nu.edu.kz (S.-W.M.)
2 Department of Civil Engineering, Faculty of Engineering, Parahyangan Catholic University, Jl. Ciumbuleuit No. 94, Bandung 40141, Indonesia; mwijaya@unpar.ac.id
3 Key Laboratory of Concrete and Prestressed Concrete Structures of Ministry of Education, Southeast University, Nanjing 210096, China; 101012332@seu.edu.cn
4 Faculty of Engineering and Informatics, University of Bradford, Bradford DB7 1DP, UK; j.h.pu1@bradford.ac.uk
* Correspondence: jong.kim@nu.edu.kz; Tel.: +7-7172-70-9136

Abstract: Tailing dams are commonly used to safely store tailings without damaging the environment. Sand tailings (also called Sediment tailings) usually have a high water content and hence undergo consolidation during their placement. As the sediment tailings are usually placed above the ground water level, the degree of saturation and permeability of the sediment tailing is associated with the unsaturated condition due to the presence of negative pore-water pressure or suction. Current practices normally focus on the analyses saturated conditions. However, this consolidation process requires the flow of water between saturated and unsaturated zones to be considered. The objective of this study is to investigate the stability and consolidation of sediment tailings for the construction of road pillars considering the water flow between saturated and unsaturated zones. The scope of this study includes the unsaturated laboratory testing of sediments and numerical analyses of the road pillar. The results show that the analyses based on saturated conditions overestimate the time required to achieve a 90% degree of consolidation. The incorporation of the unsaturated soil properties is able to optimize the design of slopes for road pillars into steeper slope angles.

Keywords: unsaturated soil; stability; consolidation; sediment

Citation: Satyanaga, A.; Wijaya, M.; Zhai, Q.; Moon, S.-W.; Pu, J.; Kim, J.R. Stability and Consolidation of Sediment Tailings Incorporating Unsaturated Soil Mechanics. *Fluids* **2021**, *6*, 423. https://doi.org/10.3390/fluids6120423

Academic Editor: Mehrdad Massoudi

Received: 21 September 2021
Accepted: 15 October 2021
Published: 24 November 2021

1. Introduction

Many major tailing dams have experienced failures in the past 30 years. Some researchers [1,2] indicate that the failure of tailing dams have contributed to catastrophic damages to environments. Several previous studies have indicated that the failures of tailing dams are normally attributed to the concurrent factors, such as: slope stability problems, bearing capacity failures, liquefactions due to earthquake and the effect of rainfall [3–8]. Some tailing dams are constructed using sand with high water content [3,9]. Hence, the sand is hydraulically placed to build the containment dam. Based on soil mechanics theory, sand as a coarse particle has high permeability under fully saturated conditions [10,11]. However, sand may have very low permeability under unsaturated conditions [12,13]. As a result, there may be some delay in the draining process of water within the sand layer during construction stages.

Sediment transport may be encountered during the water flow with high velocity within the high permeability materials. Two-phase flow is usually subjected to a complex mixture between the solid and fluid phases. It is complex to mathematically model, in particular when one considers the unsaturated conditions of the sediments [14–16]. Seepage, consolidation and stability analyses for sand tailing have been performed historically by either assuming that the soil to be completely dry, or the entire fill layer undergoing consolidation [17]. Ito and Azam [18] developed a 1-D consolidation model to estimate the behavior of sand tailing during the filling process. However, this method does not take into account the unsaturated soil mechanics principles during the construction of sand tailing. The unsaturated state of these materials may develop as a result of them being situated either above a water table or due to percolation of water down to the soil beneath the tailing because of gravity force during construction stage. In both cases, an unsaturated zone will be formed within the formation of the slope during the filling process [19]. In the unsaturated zone, the pore-water pressure within the soil becomes negative with respect to the atmospheric pressure. The negative pore-water pressure or matric suction contributes to the additional shear strength and improvement to the overall stability of the slope [20,21]. However, the permeability will drop significantly and the unsaturated zone will have very low permeability which delay the dissipation of excess pore water pressure from the saturated zone.

The objective of this study is to investigate the stability and consolidation of sand tailings road pillars constructed using the hydraulically deposited coarse sand tailings by considering the flow of water between saturated and unsaturated zones. The scope of this works includes (1) the experimental works in the laboratory to determine the saturated and the unsaturated properties of the soils and materials used in this study; (2) the assessment of the consolidation characteristics of the road pillars in order to determine the time required for full settlement to occur using Seep/W and Sigma/W [22]; (3) the determination of the design (slope angle) of road pillars to ensure the required factor of safety with respect to the additional backfilling of sediment tailings adjacent to the already constructed road pillars using Slope/W [23].

2. Saturated and Unsaturated Laboratory Testing

This study is part of the project to evaluate the feasibility of constructing road pillars using hydraulically placed coarse sand tailings as part of the rehabilitation of the Fingerboards Mineral Sands Project [24]. The road pillar is targeted to be constructed to a height of 51 m in Glenaladale, Victoria, Australia. The project lies within the Glenaladale Deposit that occurs within unconsolidated sediments and contains heavy minerals such as zircon, rutile, ilmenite and rare-earth bearing minerals (monazite and xenotime). The project represents the first phase of mineral sands production (up to 20 years) from the Glenaladale Deposit. The project area, which targets what is known as the Fingerboards resource area, is approximately 1675 ha and of this approximately 1250 ha is expected to be mined.

The construction of the road pillars is planned to use the mine sand from nearby area using only hydraulically deposited sandy tailings [24]. The index properties of the sand materials is presented in Table 1. In terms of ASTM D2487-10 [25], the classification of the coarse sand in terms of the unified soil classification system (USCS) is "SC" which stands for a clayey sand. The settling tests [26] were carried out using a sand slurry prepared at 70% solids. Two tests were used, the first with no drainage and the second with bottom drainage provided. The final settled dry densities are 1.37 and 1.40 for the tests without and with drainage, respectively. The tests show that the coarse sand settles within 4 h for bottom-drained tests. The compaction test was also performed using standard proctor method following ASTM D4253-00 [27] to obtain the maximum dry density (ρ_{dry}) and the optimum water content (w_{opt}) of the sand. The ρ_{dry} and w_{opt} are 1.465 t/m^3 and 20.5%, respectively.

The water retention curve was obtained using axis translation technique with pressure plate supported with 15 bar ceramic disc following ASTM D6838-02 [28]. The laboratory data best-fit line was fitted using the Fredlund and Xing equation [29]. The water retention curve and the fitting parameter can be seen in Figure 1. The saturated volumetric water content, the air-entry value, the residual suction and the residual volumetric water content of the sand specimen are 0.5, 7 kPa, 40 kPa and 0.05, respectively.

Table 1. Index properties of sand tailing.

Properties	
USCS	SC
Specific gravity, Gs	2.78
Gravel (%)	0
Sand (%)	69
Fines (%)	31
Bulk density (t/m3)	1.89
Water content (%)	36.4
Void ratio	0.96
Liquid Limit, LL (%)	41.6
Plastic Limit, PL (%)	19.9
Plasticity Index, PI (%)	21.7
Saturated permeability, k_s (m/s)	5.78×10^{-6}

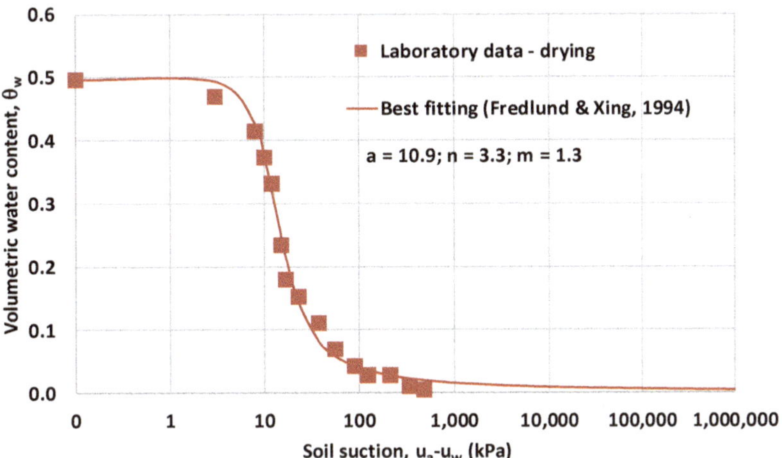

Figure 1. Water retention curve for sand.

The saturated permeability, k_s (5.78×10^{-6} m/s as shown in Table 1) was determined using a constant head laboratory method [30,31]. This is a lower permeability compared with typical poorly graded coarse sand that can range from 10^{-4} to 10^{-5} m/s [10]. The reason for this is the higher percentage of plastic fines contained in the sands of 31%. The unsaturated permeability of the sand specimen (Figure 2) was determined using statistical model following procedure presented in Satyanaga et al. [32]. The results from the oedometer test is presented in Figure 3. As the soil is hydraulically filled, the over consolidation ratio is assumed to be 1. The initial void ratio is calculated to be 0.96. The lambda (λ) and Kappa (κ) are 0.0165 and 0.0026, respectively based on the formulas shown in Equations (1) and (2).

$$\lambda = \frac{c_c}{\ln 10} \tag{1}$$

$$\kappa = \frac{c_s}{\ln 10} \tag{2}$$

In Equations (1) and (2), c_c is defined as compression index whereas c_s is defined as swelling index. In Sigma/W, it is possible to simulate the variations in the Young modulus (E') as a function of effective stress using the elastic-plastic model (w/PWP Change) [18]. The Young modulus for incorporation in Sigma/W (under effective stress material model

Elastic-Plastic) was obtained by first determining the coefficient of compressibility (m_v) from the oedometer test to determine the constrained modulus (E_{oed}) (Equation (3)).

$$E_{oed} = \frac{1}{m_v} \tag{3}$$

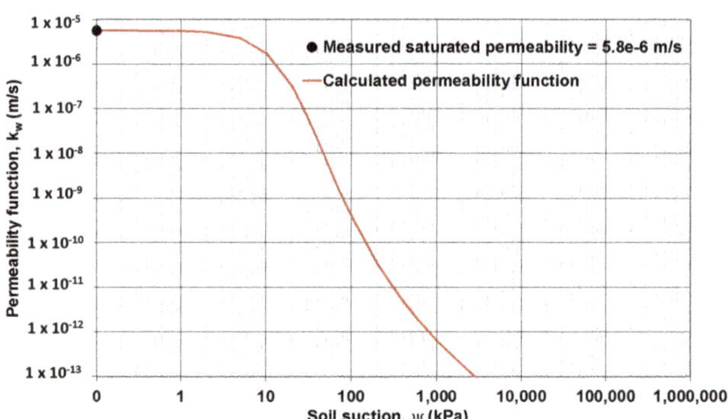

Figure 2. Permeability function of sand.

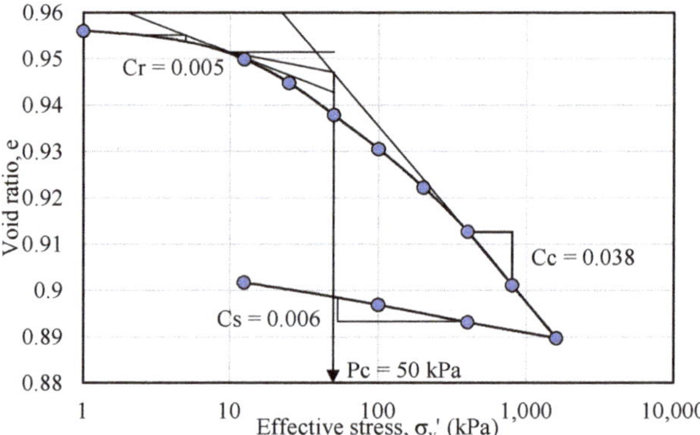

Figure 3. Consolidation curve from Oedometer testing for soil under saturated condition.

In Equation (3), m_v is defined as coefficient of compressibility. From the E_{oed}, E' can be calculated using Equation (4).

$$E' = E_{oed}\frac{(1+\nu)(1-2\nu)}{1-\nu} \tag{4}$$

One set of the saturated oedometer testing with soil suction equals to 0 (Figure 3) and three sets of the unsaturated oedometer testing under three different soil suctions (7 kPa, 17 kPa and 42 kPa) were performed on the sand specimen. For the saturated oedometer test result, the effective stress is simply calculated from the applied load (σ_v) on each loading step. Based on the unsaturated oedometer test results, the effective stress is calculated as the applied load plus the applied soil suction. A spline function is then used to represent (best fit) the four oedometer test results as shown in Figure 4. Consolidated undrained triaxial test with pore-water pressure measurement was conducted to determine the shear strength parameters of the soil. Based on the triaxial test results, the effective cohesion (c′) is 0 kPa and the effective friction angle (φ′) is 34°.

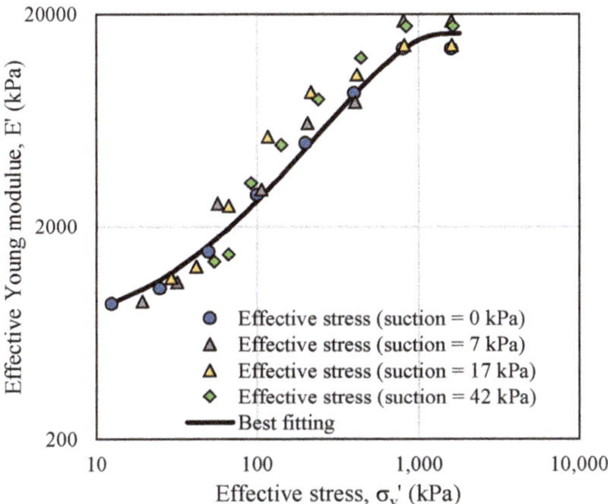

Figure 4. Change in Young Modulus (E') due to the change in effective stress.

3. Method of Analyses

The numerical analyses in this study consist of 1-D consolidation and 2-D consolidation analyses using Sigma/W, seepage analyses using Seep/W and slope stability analyses using Slope/W [33,34]. Sigma/W is a finite element software for analyzing deformation of soils whereas Slope/w is a limit equilibrium software for analyzing stability of slope. The finite element mesh had a mesh size of approximately 0.01 m in order to obtain accurate results from deformation and transient seepage analyses. The pore-water pressures variations from Seep/W were exported to Slope/w to obtain factor of safety for every time increment in the analyses. The pore-water pressures were calculated in Seep/W for every time step at each node of the finite element mesh.

1-D (coupled stress) consolidation analyses were conducted to calibrate the numerical model and parameters before proceeding with 2-D analyses; to estimate the settlement and the rate of consolidation due to the dissipation of excess pore-water pressure and due to the generation of soil suction; and to determine appropriate filling rate. Three sets of parametric studies were carried based on 1-D analyses as summarized in Table 2. The modified cam clay model was selected in the analyses since it can simulate the actual nonlinear stress-strain behavior of the soil [35].

Table 2. Parametric studies carried out based on 1-D analyses.

Case	Condition	Model	Duration of Analyses
1	Unsaturated	Modified Cam Clay model with pore-water pressure/PWP change	1 days for every 5 m filling; 10 days for the last 6 m filling; total = 19 days for 51 m filling
2	Unsaturated	Modified Cam Clay model with PWP change	Analyses were conducted until pore-water pressure change reaches equilibrium
3	Saturated	Modified Cam Clay model with PWP change	Analyses were conducted until pore-water pressure has been dissipated

In the beginning of the unsaturated analyses in Cases 1 and 2, the placement of the filling materials generated the excess pore-water pressures. As the water flowed to the in-situ soil materials, the pore-water pressure dissipated and slowly became negative as the top of fill material was located above the ground water table which is assumed to be at the ground level. The settlement was then contributed due to consolidation settlement (when the excess pore-pressure dissipated to 0 kPa) and shrinkage settlement (when the pore-water pressure goes below 0 kPa). In Sigma/W, both consolidation and shrinkage settlement were calculated based on effective stress principle. The saturated analysis in case 3 was carried out to understand the contribution of soil suction in determining the rate of settlement of the filling materials.

The numerical model only allowed instantaneous placement of filling materials. Therefore, the road pillar model (51 m high) was divided into 10 layers to understand the characteristics of pore-water pressure changes with time for every 5 m filling materials. Two different scenarios were used to determine the time step for every 5 m filling materials. In the first scenario, 1 day was required to complete every 5 m filling process except it is required 10 days for the last filling process. In the second scenario, every 5 m filling materials were consolidated until the pore-water pressure reached hydrostatic condition. Settlement analyses were carried out by first obtaining the settlement at the node located at the top of the filling materials at each stage. The total settlement is equal to the summation of the settlement obtained from the top node at each filling stage (the settlement of the node after the next filling stage was ignored).

The 2-D deformation and stability analyses were conducted with the following objectives: to investigate the effect of slope angle on the settlement of the road pillar and to analyse the stability of the road pillar under different slope angle. The width of the crest of the road pillar is maintained at 63 m for different inclinations of road pillar. This study focuses on the construction of road pillar with the following slope angles: 25°and 30°. For both geometries, the stability analyses were performed with the assumption of the presences of tailing materials at one side of the road pillar.

4. One Dimensional (1-D) Numerical Analyses

The soil model was divided into 11 regions. The first region represented the in-situ soil material while the other 10 regions represented the filling materials. Every region had 5 m height except the last region had 6 m height. Thus, the total height of the filling materials was 51 m as shown in Figure 5a. The filling materials were placed in stages, on top of each other and the whole soil column was constrained laterally at the side while it was fixed at the bottom. The in-situ soil consists of gravel layer with Young modulus (E') equal to 200 MPa as obtained from Oedometer testing data [24] and high permeability. Therefore, no excess pore-water pressure was generated in the in-situ soil.

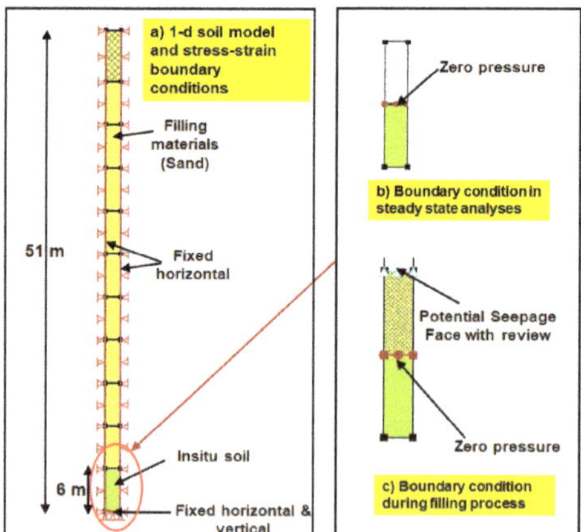

Figure 5. Soil model and boundary condition.

Steady state condition was first generated using Seep/W based on steady state analyses. In the steady state analyses, zero pressure boundary condition (pressure head equals to 0 kPa) was simulated at the top of the in-situ soil (Figure 5b). Afterward, in-situ analyses in Sigma/W were conducted to generate the initial stress of the in-situ soil. Coupled stress analysis was conducted using Sigma/W to simulate the hydraulic filling process. As the hydraulic boundary condition, potential seepage face with review (to simulate the runoff on the surface of the soil) the was applied at the top of the soil layer while zero pressure was applied at the top of the in-situ soil as shown in Figure 5c. In the saturated analyses under case 3, the hydraulic boundary condition for steady state analyses was the same as the unsaturated analyses in cases 1 and 2. However, zero pressure boundary condition was applied at the top of the filling material (instead of seepage review) and at the top of the in-situ soil during the filling stage.

In numerical analyses of case 1, a 1-day time step was applied on the simulation of each fill layer while 10- day time steps was applied on the simulation of the last fill layer. This is to ensure that the total settlement has reached equilibrium condition. The total settlement of the entire fill materials at each step is shown in Figure 6. It is shown that at the end of the filling stage, the settlement is 2.68 m with the final distribution of pore-water pressure is shown in Figure 7. It can be seen that the excess pore-water pressure near the surface (within 40–50 m) has dissipated quickly since the negative pore-water pressure was developed within this layer due to the flow of water into the deeper layer.

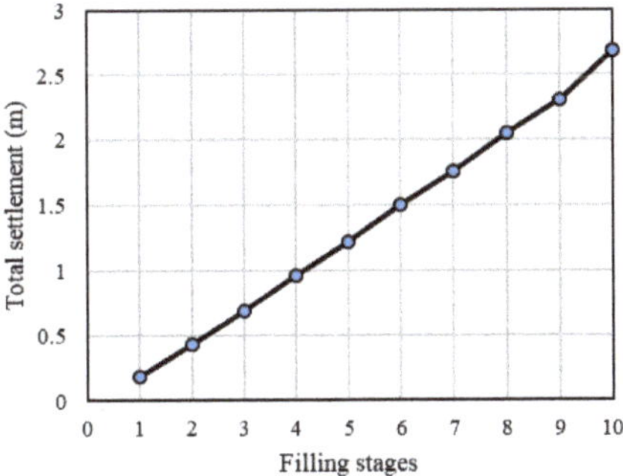

Figure 6. Total settlement of the fill materials for case 1.

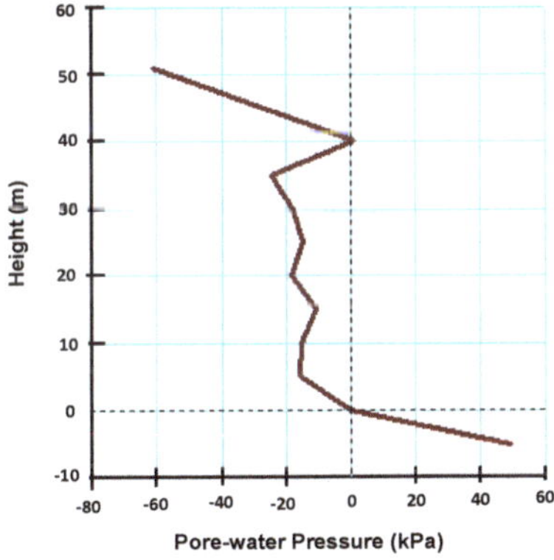

Figure 7. Final distribution of pore-water pressure at the last filling stage for case 1.

Numerical analysis in case 2 was carried out to determine the required time for the negative pore-water pressure to reach the equilibrium condition. The node under scrutiny is the top node of the fill as this node has the highest negative pore-water pressure. The degree of consolidation (U_z) is commonly used to determine appropriate time to determine when the consolidation has been completed (Equation (5)).

$$U_z(\%) = \left(1 - \frac{\Delta u_w}{\Delta u_0}\right)100\% \tag{5}$$

where Δu_0 is the initial excess pore-water pressure due to the applied load, Δu_w is the current excess pore-water pressure. The applied load always caused the pore-water pressure at the top node to reach zero and then slowly went to pore-water pressure at equilibrium/hydrostatic condition (u_{eq}) (Equation (6)).

$$u_{eq} = (GWL - z)\gamma_w \tag{6}$$

where GWL was the ground water level which was set to zero in the numerical model and γ_w is the unit weight of water (9.81 kN/m^3). Δu_0 was defined for the node located at the top of the fill (Equation (7)). Δu_w can be defined using Equation (8).

$$\Delta u_0 = u_0 - u_{eq} \tag{7}$$

$$\Delta u_w = u_w - u_{eq} \tag{8}$$

where u_0 is the pore-water pressure after application of load which was always equals to zero for the top node while u_w is the pore-water pressure for the node which was changing with time.

Figure 8 shows the example of changes in pore-water pressure of the top node on different filling stage. It can be seen that a reasonable waiting time can be achieved for 0–5 m fill (95 days to reach U_z = 95%) and 5–10 m fill (2.5 years to reach U_z = 95%). However, for 10–15 m fill, it required 10 years to reach Uz = 80% and 32 years to reach U_z = 95%. At the final fill layer (45–51 m), it requires 1600 years to reach U_z = 80% and 3400 years to reach U_z = 95%. Based on this analysis, it was not reasonable to wait for the pore-water pressure to reach equilibrium prior to applying the next load.

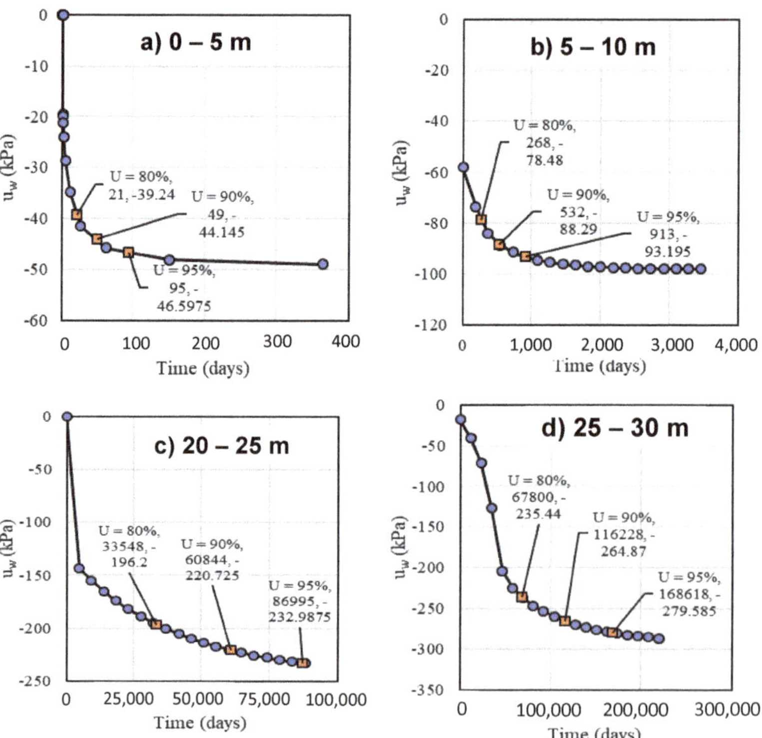

Figure 8. Time required for the negative pore-water pressure to reach equilibrium in each filling stage.

The settlement from case 2 can be considered as the final total settlement, the average degree of consolidation (U_{ave}) can be calculated using Equation (9).

$$U_{ave}(\%) = \frac{s_{case1}}{s_{case2}} \tag{9}$$

where s_{case1} is the total settlement at case 1 while s_{case2} is the total settlement at case 2. The U_{ave} of case 1 is 85%.

In order to fully dissipate all excess pore-water pressures, every filling stage duration in case 3 was extended for two days. Figure 9 shows the comparison of total settlement between cases 1, 2 and case 3. It can be seen that case 3 has the least settlement due to ignoring the effect of negative pore-water pressure which contributing to the additional settlement. The comparisons of analyses time for cases 1, 2 and 3 are presented in Table 3. Based on the three simulation results from 1-D consolidation analyses, the model used in the unsaturated analyses of case 1 was used in 2-D numerical analysis. In addition, the filling stage was decided to be 5 m/stage with every stage being completed in 1 day.

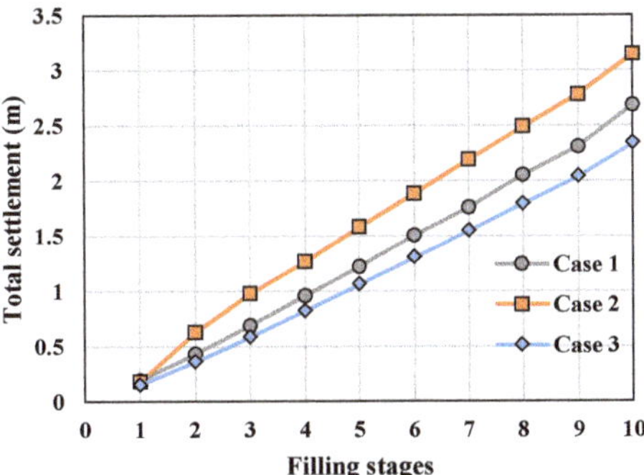

Figure 9. Comparison of total settlement from analyses of case 1, case 2 and case 3.

Table 3. Analyses time for each stage in Case 1, Case 2 and Case 3.

Case	1	2	3	4	5	6	7	8	9	10
Case 1 (day)	1	1	1	1	1	1	1	1	1	10
Case 2 (year)	1	9.5	90	190	241	602	1506	2410	3012	6025
Case 3 (day)	2	2	2	2	2	2	2	2	2	2

2-D Numerical Analysis

Figure 10 shows the 2-D finite element model of the road pillar. Stability analysis of the 2-D model was carried using simplified Bishop model in Slope/W. No tension crack was considered, and the slip surface was determined based on grid and radius method [11]. The results of the consolidation and stability analyses of the road pillar with different inclinations are presented in this section. The slope stability analysis was carried out for every time step to consider the effect of PWP dissipation in the stability of the fill layer. This analysis models the road pillar in both X and Z directions and seepage can flow in two directions, vertically and laterally. For each of the three geometries, three scenarios were considered in the 2-D stability analyses. In scenario 1, the stability analyses were carried out to investigate the stability of road pillar only. In scenario 2, the stability analyses were conducted to study the stability of road pillar with the assumption of the presences of tailing materials at one side of the road pillar. In scenario 3, the stability analyses were performed to investigate the stability of road pillar with the assumption of the presence of 4.5 m high of water at one side of the road pillar. The numerical models for scenario 1, scenario 2 and scenario 3 are presented in Figure 10.

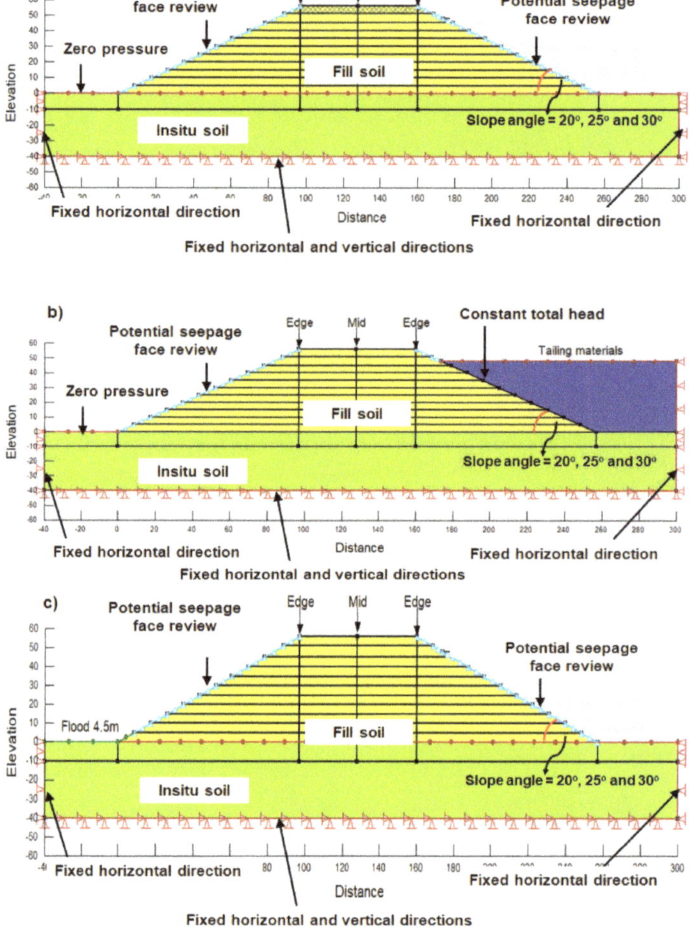

Figure 10. 2-D finite element model for analyses of (**a**) scenario 1; (**b**) scenario 2 and (**c**) scenario 3.

Consolidation settlements at the midpoint and edge of the model based on scenario 1 (Figure 10a) were compared with the results from the 1-D consolidation analyses. The finite element model was separated into the in-situ soil and the filling material. The filling material soil was separated into 11 regions which are 0–5 m, 5–10 m, 10–15 m, 15–20 m, 20-25 m, 25–30 m, 30–35 m, 35–40 m, 40–45 m, 45–51 m and 51–56 m. The soil model and properties of in-situ soil and fill soil followed those used as input in 1-D consolidation analysis of case 1. Every filling layer was consolidated for 1 day with 10 time steps and 2.4 h for each step. The in-situ soil was fixed at horizontal and vertical directions as shown in Figure 10. The zero pressure boundary condition was applied at the interface between the in-situ soil and the fill soil while the potential seepage face with review was applied at the slope surface of the fill layer as shown in Figure 10.

The total settlement at the end of the construction located at the edge and the mid of the crest for the three geometries are summarized in Table 4. The results present that the differential settlement is around ±0.01m for all cases. The settlement at the middle and the edge of the crest of the fill layer for the three geometries were compared with the one from 1-D analysis as shown in Figure 11. The results of the 1-D and 2-D consolidation settlement analyses are comparable as they follow similar trends. In addition, all the final pore-water pressures at the middle of the model at the end of stage 11 or after 56 m filling have become negative for all geometries (Figure 12).

Table 4. Summary of the total settlement from 2-D numerical analyses.

Case	Mid (m)	Edge (m)
Geometry 1 (20°)	3.13	3.13
Geometry 2 (25°)	3.13	3.13
Geometry 3 (30°)	3.13	3.15

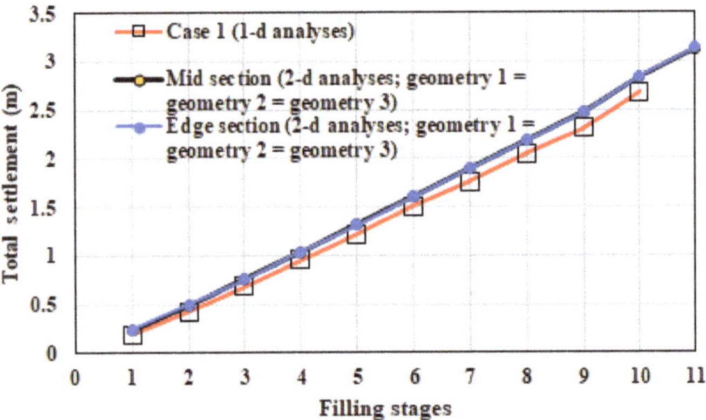

Figure 11. Comparisons of total settlement from 1-d analyses (based on Case 1) and 2-d analyses (based on three different geometries).

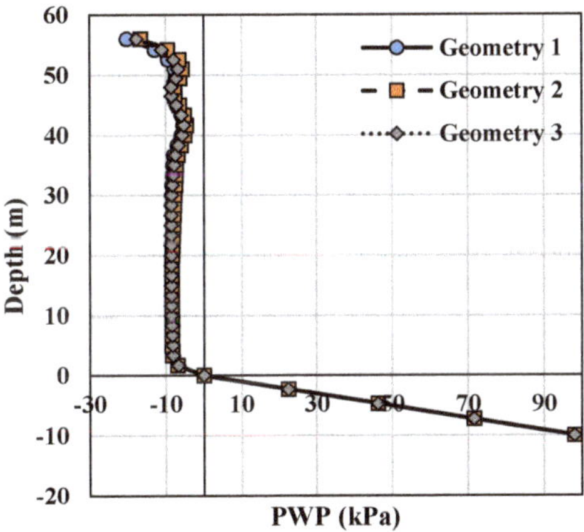

Figure 12. Pore-water pressure profiles at the end of stage 11 for geometry 1, geometry 2 and geometry 3.

The results from stability analyses of scenario 1 are presented in Figure 13. It is shown that the critical condition is observed when the fill layer has just been placed. As the pore-water pressure dissipated, the factor of safety slowly increases. It is shown that geometry 3 has lower factor of safety compare to other geometries. The lowest factor of safety is observed when the first fill is just placed (0–5 m) with factor of safety = 1.15. The factor of safety increases to 1.73 within 2.4 h. For the rest of the simulations, all factor of safety values are higher than 1.3. The final factor of safety of the road pillar is 1.344 which considered to be the minimum in terms of the design acceptance criteria [36].

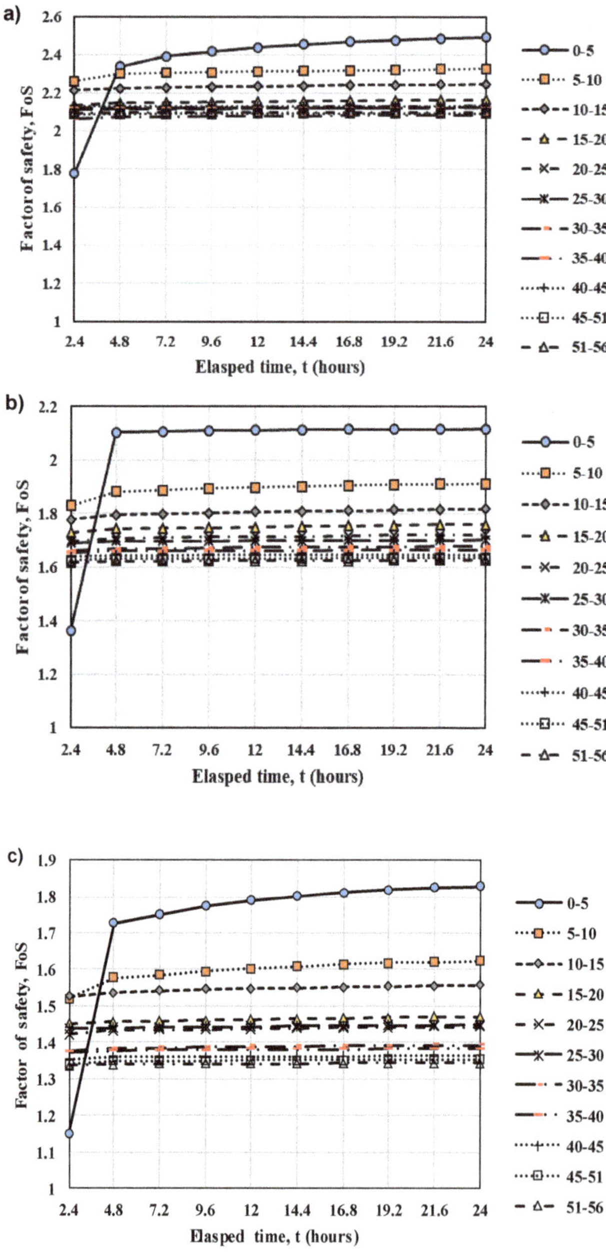

Figure 13. Variations of factor of safety with time based on stability analyses of scenario 1 for (**a**) geometry 1; (**b**) geometry 2 and (**c**) geometry 3.

In the stability analyses of scenario 2, the tailing sediment was assumed to be present at one side of the embankment. Given that the tailing material was assumed the same as the road pillar material, the tailing material was considered to be instantaneously placed at one side of the road pillar and then consolidated for 1 year based on 12 time steps. Constant total head was applied at the boundary between road pillar and tailing materials to model the groundwater table in the numerical analysis (Figure 10). The stability of the opposite side of road pillar is checked using the same method used for checking the stability of 56 m of road pillar only. In this model, the pore-water pressure of the tailing sediment can be negative as shown in Figure 14. Essentially, the backfilled tailings is allowed to dewater and therefore both the road pillar and the tailings becomes unsaturated. Figure 15 shows the variations in safety factor with time for scenario 2. The difference was not significant over 1 year duration and the most critical condition was generated from the stability analysis of slope with 30° slope angle with factor of safety is about 1.34 to 1.35.

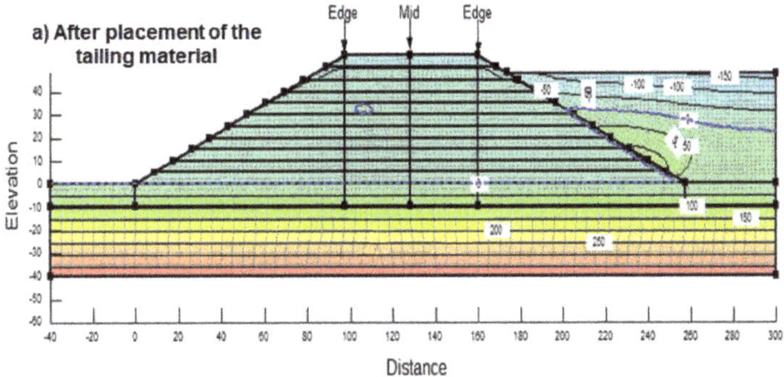

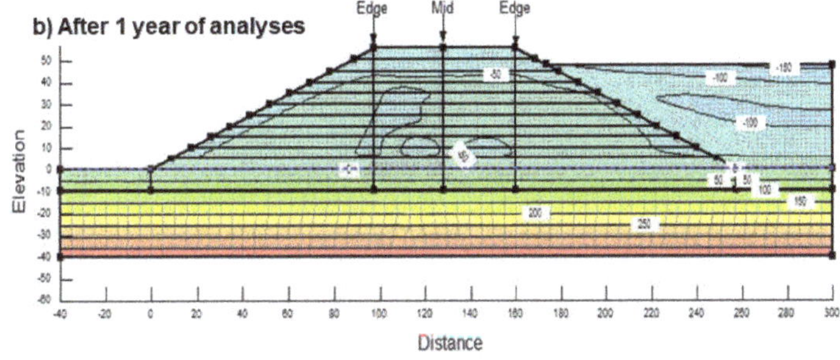

Figure 14. Distribution of pore-water pressures due to the placement of tailing material without zero pressure boundary condition applied on top of the tailing material (Approach 2).

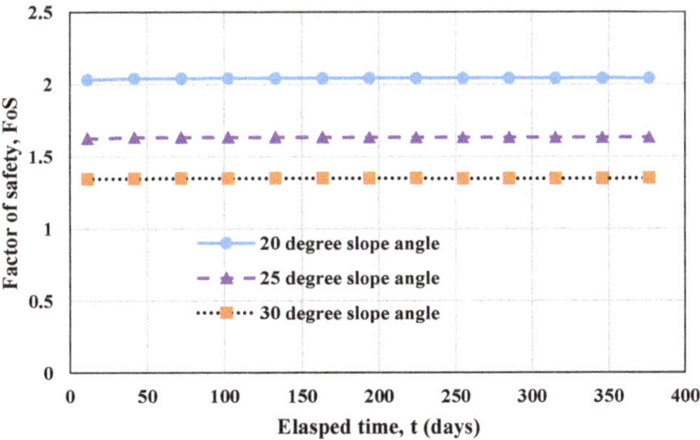

Figure 15. Variations of factor of safety with time.

In the anticipation of climate change, the additional stability analyses under extreme rainfall conditions were carried out. A 4.5 m water flooding above the pit floor event is assumed at one side of road pillar in the model. This rainwater is sufficient to model a 1% annual exceedance probability (AEP) 72 h rainfall event (421 mm of rainfall) on a slope of 5% with a bund at the toe and 100% runoff coefficient [24]. A constant total head of 4.5 m is applied at one side of the road pillar over 1 year duration. Figure 16 shows the variations in factor of safety based on the stability analyses for Scenario 3. Geometry 3 (30° slope angle) shows the most critical condition which has a factor of safety around 1.35.

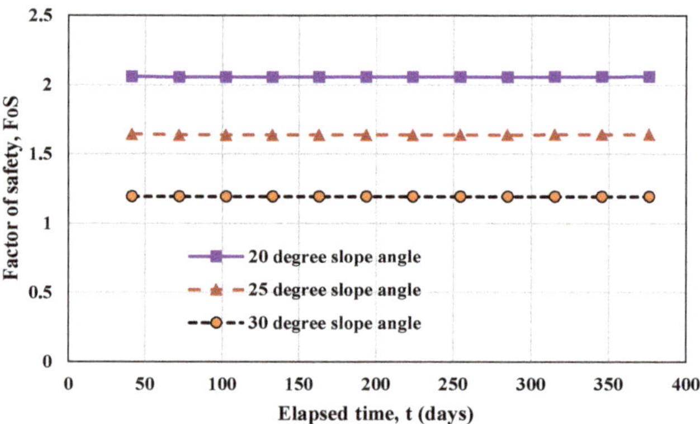

Figure 16. Variations of factor of safety with time based on stability analyses for scenario 3.

Based on the numerical analyses results, it is shown that ignoring the effect of negative pore-water pressure will underestimate the settlement of the fill materials. However, it is not reasonable to wait for the entire pore-water pressures to reach equilibrium since it took a very long time (>6000 years). In addition, based on case 1 simulation, 1 day waiting time for each of the filling stage was sufficient for the degree of consolidation to reach 85% and thus the model based on case 1 was recommended. The filling of 51 m contributes to a total settlement of 3.1 m. Hence, the final elevation of the road pillar based on 1-D analyses is 51 − 3.1 m = 47.9 m.

The results from stability analyses indicate that the incorporation of unsaturated soil properties help to optimize the design of slope. The slope still has high factor of safety although the inclination is increased to 30° slope angle which has been evaluated under three different conditions, without rainfall and tailing sediment, with tailing sediments and with rainfall loading. The lowest factor of safety of slope with 30° slope angle is around 1.35 under maximum rainfall loading.

It can be seen the factor of safety of 30° slope angle (1.35) in Figure 16 is lower than that from Figure 13c (1.82). The factor of safety of of 25° slope angle (1.70) in Figure 16 is lower than that from Figure 13b (2.13). The factor of safety of of 20° slope angle (2.10) in Figure 16 is lower than that from Figure 13b (2.50). The comparisons of factor of safety between Figures 13 and 16 indicate the importance of maintaining the unsaturated conditions of the slope. If there is potential of flooding next to the road pillar, it is recommended to provide the appropriate drainage within the structure itself, such as: horizontal drains; geodrain and surface drain to ensure the water can be drained out properly in the case flooding occur due to the effect of climate change.

5. Conclusions

The key conclusions can be summarized as follows:

- Laboratory testing has shown that the coarse sand tailings has a 34% (by mass) of medium plastic fines (<0.075 mm), This is higher than conventional coarse sand tailings and has resulted in the sand having a lower permeability and lower shear strength then normal poorly graded coarse sand.

- Numerical consolidation modelling using both 1-D and 2-D showed that despite the higher fines content, the hydraulically deposited coarse sand tailings will be able to consolidate very rapidly and that a degree of consolidation of 85% is achievable within days. The results from numerical analyses also indicated that it is important to incorporate the unsaturated soil mechanics principles in the consolidation analyses. The ignorance of the unsaturated soil mechanics principles will generate less settlement in the consolidation analyses.
- In terms of the stability modelling, the minimum FOS for embankment slope angles up to and including 30° is 1.35. This value is higher than minimum factor of safety of 1. It shows that the hydraulically placed method to stack the fill materials in the road pillars naturally can achieve a higher slope angle if the unsaturated soil properties is included in the analyses.
- The appropriate drainage should be provided within the embankment to maintain the unsaturated condition and to ensure the stability of the embankment under heavy rainfall loadings.

Author Contributions: A.S., S.-W.M. and J.R.K. conceptualized the study; A.S., M.W. and Q.Z. implemented data processing under the supervision of J.R.K. and S.-W.M.; the original draft of the manuscript was written by A.S., M.W. and Q.Z. with editorial contributions from S.-W.M., J.P. and J.R.K.; the funding acquisition was made by J.R.K. All authors have read and agreed to the published version of the manuscript.

Funding: This research was funded by Nazarbayev University, grant number SOE2017004. The authors are grateful for this support. Any opinions, findings, and conclusions or recommendations expressed in this material are those of the author(s) and do not necessarily reflect the views of the Nazarbayev University.

Acknowledgments: The Authors would like to express sincere gratitude for the tremendous advice from Harianto Rahardjo and outstanding assistance from Ken Mercer for the opportunity to obtain the experimental data for the study.

Conflicts of Interest: The authors declare no conflict of interest.

References

1. Liu, R.; Liu, J.; Zhang, Z.; Borthwick, A.; Zhang, K. Accidental water pollution risk analysis of mine tailings ponds in Guanting reservoir Watershed, Zhangjiakou city, China. *Int. J. Environ. Res. Public Health* **2015**, *12*, 15269–15284. [CrossRef] [PubMed]
2. Vogel, A. Failures of dams—Challenges to the present and the future. In *IABSE Symposium Report*; International Association for Bridge and Structural Engineering: Zürich, Switzerland, 2013; pp. 178–185. [CrossRef]
3. Azam, S.; Li, Q. Tailings dam failures: A review of the last one hundred years. *Geotech. News* **2010**, *28*, 50–54.
4. Rico, M.; Benito, G.; Salgueiro, A.R.; Díez-Herrero, A.; Pereira, H.G. Reported tailings dam failures: A review of the European incidents in the worldwide context. *J. Hazard Mater.* **2008**, *152*, 846–852. [CrossRef] [PubMed]
5. Glotov, V.E.; Chlachula, J.; Glotova, L.P.; Little, E. Causes and environmental impact of the gold-tailings dam failure at Karamken, the Russian Far East. *Eng. Geol.* **2018**, *245*, 236–247. [CrossRef]
6. Cambridge, M.; Shaw, D. Preliminary reflections on the failure of the Brumadinho tailings dam in January 2019. *Dams Reserv.* **2019**, *29*, 13–123. [CrossRef]
7. Villavicencio, G.; Espinace, R.; Palma, J.; Fourie, A.; Valenzuela, L. Failures of sand tailings dams in a highly seismic country. *Can. Geotech. J.* **2014**, *51*, 449–464. [CrossRef]
8. Wei, Z.; Yin, G.; Wang, J.; Wan, L.; Li, G. Design, construction and management of tailings storage facilities for surface disposal in China: Case studies of failures. *J. Waste Manag. Res.* **2013**, *31*, 106–112. [CrossRef]
9. Chen, R.; Ding, X.; Ramey, D.; Lee, I.; Zhang, L. Experimental and numerical investigation into surface strength of mine tailings after biopolymer stabilization. *Acta Geotech.* **2016**, *11*, 1075–1085. [CrossRef]
10. Holtz, R.D.; Kovacs, W.D.; Sheahan, T.C. *An Introduction to Geotechnical Engineering*; Pearson: New York, NY, USA, 2011; p. 864. ISBN 9780132496346.
11. Satyanaga, A.; Rahardjo, H.; Hua, C.J. Numerical simulation of capillary barrier system under rainfall infiltration. *ISSMGE Int. J. Geoengin. Case Hist.* **2019**, *5*, 43–54. [CrossRef]
12. Fredlund, D.G.; Rahardjo, H. *Soil Mechanics for Unsaturated Soil*; John Wiley: New York, NY, USA, 1993; p. 517. [CrossRef]

13. Rahardjo, H.; Kim, Y.; Gofar, N.; Satyanaga, A. Analyses and design of steep slope with GeoBarrier System under heavy rainfall. *Geotext. Geomembr.* **2020**, *48*, 157–169. [CrossRef]
14. Pu, J.H.; Wallwork, J.T.; Khan, M.A.; Pandey, M.; Pourshahbaz, H.; Satyanaga, A.; Hanmaiahgari, P.R.; Gough, T. Flood suspended sediment transport: Combined modelling from dilute to hyper-concentrated flow. *Water* **2021**, *13*, 379. [CrossRef]
15. Pu, J.H.; Hussain, K.; Shao, S.-D.; Huang, Y.-F. Shallow sediment transport flow computation using time-varying sediment adaptation length. *Int. J. Sediment Res.* **2014**, *29*, 171–183. [CrossRef]
16. Pu, J.H.; Huang, Y.; Shao, S.; Hussain, K. Three-gorges dam fine sediment pollutant transport: Turbulence SPH model simulation of multi-fluid flows. *J. Appl. Fluid Mech.* **2016**, *9*, 1–10. [CrossRef]
17. Liu, K.; Yin, J.H.; Chen, W.B.; Feng, W.Q.; Zhou, C. The stress–strain behaviour and critical state parameters of an unsaturated granular fill material under different suctions. *Acta Geotech.* **2020**, *15*, 3383–3398. [CrossRef]
18. Rodríguez, R.; Muñoz-Moreno, A.; Caparrós, A.V.; García-García, C.; Brime-Barrios, A.; Arranz-González, J.C.; Rodríguez-Gómez, V.; Fernández-Naranjo, F.J.; Alcolea, A. How to prevent flow failures in tailings dams. *Mine Water Environ.* **2021**, *40*, 83–112. [CrossRef]
19. Ito, M.; Azam, S. Large-strain consolidation modeling of mine waste tailings. *Environ. Syst. Res.* **2013**, *2*, 7. [CrossRef]
20. Rahardjo, H.; Kim, Y.; Satyanaga, A. Role of unsaturated soil mechanics in geotechnical engineering. *Int. J. Geo-Eng.* **2019**, *10*, 1–23. [CrossRef]
21. Satyanaga, A.; Rahardjo, H. Role of unsaturated soil properties in the development of slope susceptibility map. *Proc. Inst. Civ. Eng. Geotech. Eng.* **2020**. [CrossRef]
22. Rahardjo, H.; Satyanaga, A.; Leong, E.C.; Ng, Y.S.; Foo, M.D.; Wang, C.L. Slope failures in Singapore due to rainfall. In Proceedings of the 10th Australia New Zealand Conference on Geomechanics "Common Ground", Brisbane, Australia, 21–24 October 2007; Volume 2, pp. 704–709.
23. Rahardjo, H.; Satyanaga, A.; Harnas, F.R.; Leong, E.C. Use of dual capillary barrier as cover system for a sanitary landfill in Singapore. *Indian Geotech. J.* **2016**, *46*, 228–238. [CrossRef]
24. Kalbar Resources Ltd. *Evaluation of the Stability and Consolidation of Hydraulically Placed Coarse Sand Tailings Road Pillars*; Kalbar Resources Ltd.: Bairnsdale, Australia, 2020; p. 69.
25. ASTM D2487-17. *Standard Practice for Classification of Soils for Engineering Purposes (Unified Soil Classification System)*; ASTM International: West Conshohocken, PA, USA, 2017.
26. ASTM D4254-16. *Standard Test Methods for Minimum Index Density and Unit Weight of Soils and Calculation of Relative Density*; ASTM International: West Conshohocken, PA, USA, 2016.
27. ASTM D4253-00. *Standard Test Methods for Maximum Index Density and Unit Weight of Soils Using a Vibratory Table*; ASTM International: West Conshohocken, PA, USA, 2006.
28. ASTM D6838-02. *Standard Test Methods for the Soil-Water Characteristic Curve for Desorption Using Hanging Column, Pressure Extractor, Chilled Mirror Hygrometer, or Centrifuge*; ASTM International: West Conshohocken, PA, USA, 2008.
29. Rahardjo, H.; Satyanaga, A.; Mohamed, H.; Ip, S.C.Y.; Rishi, S.S. Comparison of soil-water characteristic curves from conventional testing and combination of small-scale centrifuge and dew point methods. *J. Geotech. Geol. Eng.* **2019**, *37*, 659–672. [CrossRef]
30. Head, K.H. *Manual of Soil Laboratory Testing*; Pentech Press: London, UK, 1986.
31. Zhai, Q.; Rahardjo, H.; Satyanaga, A. A pore-size distribution function based method for estimation of hydraulic properties of sandy soils. *Eng. Geol.* **2018**, *246*, 288–292. [CrossRef]
32. Satyanaga, A.; Rahardjo, H.; Koh, Z.H.; Mohamed, H. Measurement of a soil-water characteristic curve and unsaturated permeability using the evaporation method and the chilled-mirror method. *J. Zhejiang Univ. Sci. A* **2019**, *20*, 368–375. [CrossRef]
33. Paul, S.; Oppelstrup, J.; Thunvik, R.; Magero, J.M.; Ddumba Walakira, D.; Cvetkovic, V. Bathymetry development and flow analyses using two-dimensional numerical modeling approach for Lake Victoria. *Fluids* **2019**, *4*, 182. [CrossRef]
34. Liu, E.; Yu, H.S.; Deng, G.; Zhang, J.; He, S. Numerical analysis of seepage–deformation in unsaturated soils. *Acta Geotech.* **2014**, *9*, 1045–1058. [CrossRef]
35. Qubain, B.S.; Li, J.; Change, K.E. Coupled consolidation analysis of field instrumented preloading program. *J. Geotech. Geonvironmental Eng.* **2014**, *140*, 04013048. [CrossRef]
36. Department of Transport and Main Roads, Queensland, Australia. *Manual for Geotechnical Design Standard—Minimum Requirements*; Department of Transport and Main Roads: Queensland, Australia, 2020; p. 44.

Article

Toward a Better Understanding of Sediment Dynamics as a Basis for Maintenance Dredging in Nagan Raya Port, Indonesia

Muhammad Zikra *, Shaskya Salsabila and Kriyo Sambodho

Ocean Engineering Department, Institut Teknologi Sepuluh Nopember, Surabaya 60111, Indonesia; shaskyasalsa2@gmail.com (S.S.); dhodhot@gmail.com (K.S.)
* Correspondence: mzikro@gmail.com

Abstract: The Port of 2 × 110 MW Nagan Raya Coal Fired Steam Power Plant is one of the facilities constructed by the State Electricity Company in Aceh Province, Indonesia. During its operation, which began in 2013, the port has dealt with large amounts of sedimentation within the port and ship entrances. The goal of this study is to mitigate the sedimentation problem in the Nagan Raya port by evaluating the effect of maintenance dredging. Field measurements, and hydrodynamic and sediment transport modeling analysis, were conducted during this study. Evaluation of the wind data showed that the dominant wind direction is from south to west. Based on the analysis of the wave data, the dominant wave direction is from the south to the west. Therefore, the wave-induced currents in the surf zone were from south to north. Based on the analysis of longshore sediment transport, the supply of sediments to Nagan Raya port was estimated to be around 40,000–60,000 m^3 per year. Results from the sediment model showed that sedimentation of up to 1 m was captured in areas of the inlet channel of Nagan Raya port. The use of a passing system for sand is one of the sedimentation management solutions proposed in this study. The dredged sediment material around the navigation channel was dumped in a dumping area in the middle of the sea at a depth of 11 m, with a distance of 1.5 km from the shoreline. To obtain a greater maximum result, the material disposal distance should be dumped further away, at least at a depth of 20 m or a distance of 20 miles from the coastline.

Keywords: wave; current; sediment; maintenance dredging; Nagan Raya

Citation: Zikra, M.; Salsabila, S.; Sambodho, K. Toward a Better Understanding of Sediment Dynamics as a Basis for Maintenance Dredging in Nagan Raya Port, Indonesia. *Fluids* **2021**, *6*, 397. https://doi.org/10.3390/fluids6110397

Academic Editors: Jaan H. Pu and Mehrdad Massoudi

Received: 14 July 2021
Accepted: 29 October 2021
Published: 3 November 2021

Publisher's Note: MDPI stays neutral with regard to jurisdictional claims in published maps and institutional affiliations.

1. Introduction

As the first Coal Fired Steam Power Plant present in Aceh Province, the existence of this power plant operation is very important for the local community as it supports the electricity demand in the south-west coast of Aceh Province. However, since the completion of the port construction process with the breakwater in 2013, there was a sedimentation process inside the port and the mouth of channel [1]. The breakwater structure was extended and some groins were also constructed [2,3] to minimize the problem, but these solutions have not been maximally effective in protecting the port basin from sedimentation. Sedimentation threatened the operation of the Nagan Raya port and steam power plant. A suitable port-basin depth for the anchoring of coal supply vessels is no longer achieved. In addition, the inlet channel for the water supply for the cooling of the power plant engine placed in the harbor basin is experiencing siltation. This siltation led to a reduction and decrease in the quality of the condenser coolant water supply, thus disrupting the productivity and service of the power plant for the people of Aceh Darussalam.

To overcome these problems, dredging activity was carried out to maintain the navigation channel depth and the port basin. However, the dredging solution is temporary and a more permanent solution is required to solve the sedimentation problem at the port. Because routine dredging activity has significantly increased costs and can lead to port downtimes, dredging optimization is strongly encouraged. On the other hand, the volume of sediment will continue to increase if the main problems of the sedimentation process are not addressed.

In short, over the years, new technologies have been developed as alternative methods for dredging. The first alternative was the use of fixed coastal structures such as groins [4,5], sand traps [6,7] and current deflection walls [8,9] to remove or reduce the amount of sediment from the port entrance. Recently, C. Juez et al. (2018) analyzed the effect of bank macro-roughness on sediment transportation [10]. Their laboratory study showed a clear function of the macro-roughness elements in terms of the trapping of fine sediments in the channel.

The second alternative to be developed was the use of water injections to lift and separate the grains of sediment from the seabed. The first design of a method for this system was proposed by Weisman et al. (1996) [11]. Water is pumped into a perforated buried pipeline below the seafloor for seabed maintenance. The third technological approach to be introduced is known as the sand bypassing system. Since the early 1930s, the sand bypassing system has been used in the United States [12]. Sand bypassing has a small environmental impact, but has high installation costs with uncertain operational costs [13]. Furthermore, the sand bypassing system usually works on a continuous basis and the system is able to ensure the prevention of sedimentation over time for the maintenance of the seabed.

The purpose of this study was to mitigate the sedimentation problem in Nagan Raya port by evaluating the effect of dredging maintenance activities. This study was mainly achieved in two parts. First, field measurements were conducted to identify the environmental characteristics of the study area, such as the wave climate, sediment properties, tide, current speed, wind climate, sediment transport volume and bathymetry. Secondly, numerical models (ShorelineS and Delft3D) were applied to provide an overview of the hydrodynamic features occurring within ports that directly or indirectly lead to sedimentation. The process-based Delft3D model was used to determine the sediment sources and sink that corresponded to areas where the sediment was disposed or distributed within the field of study. This Delft3D model has been implemented and applied in various coastal areas around the world to explore the impact of dredging material on sedimentation, e.g., in [14–17].

2. Study Area

The Nagan Raya Coal Fired Steam Power Plant (known as PLTU Nagan Raya) is a power plant that uses coal as its fuel and can supply electricity at a capacity of 2×110 MW. This Steam Power Plant is located in Gempong Sauk Puntong Village, District of Nagan Raya, Nanggroe Aceh Darussalam Province, Indonesia. The port is one of the facilities that supports the main operation of the coal-fired power plant in supplying coal as the main fuel for engines used in driving. The port is geographically located at coordinates $04^\circ 06.116'$ North latitude and $96^\circ 11.6'$ East longitude as shown in Figure 1.

In the north part, there is the Krueng Meureubo river, which is located about 6 km from the location of Nagan Raya port. Meanwhile, 7 km south from Nagan Raya port, there is the Krueng Nagan river. Both rivers have a large watershed and carry significant sediment to the river mouth. Because of the long distance and the influence of longshore currents that cuts the movement of sediment, the sediment contribution originating from the river is considered non-existent. In this study, only sediments from around the coastal area of Nagan Raya power plant were considered in the numerical simulation.

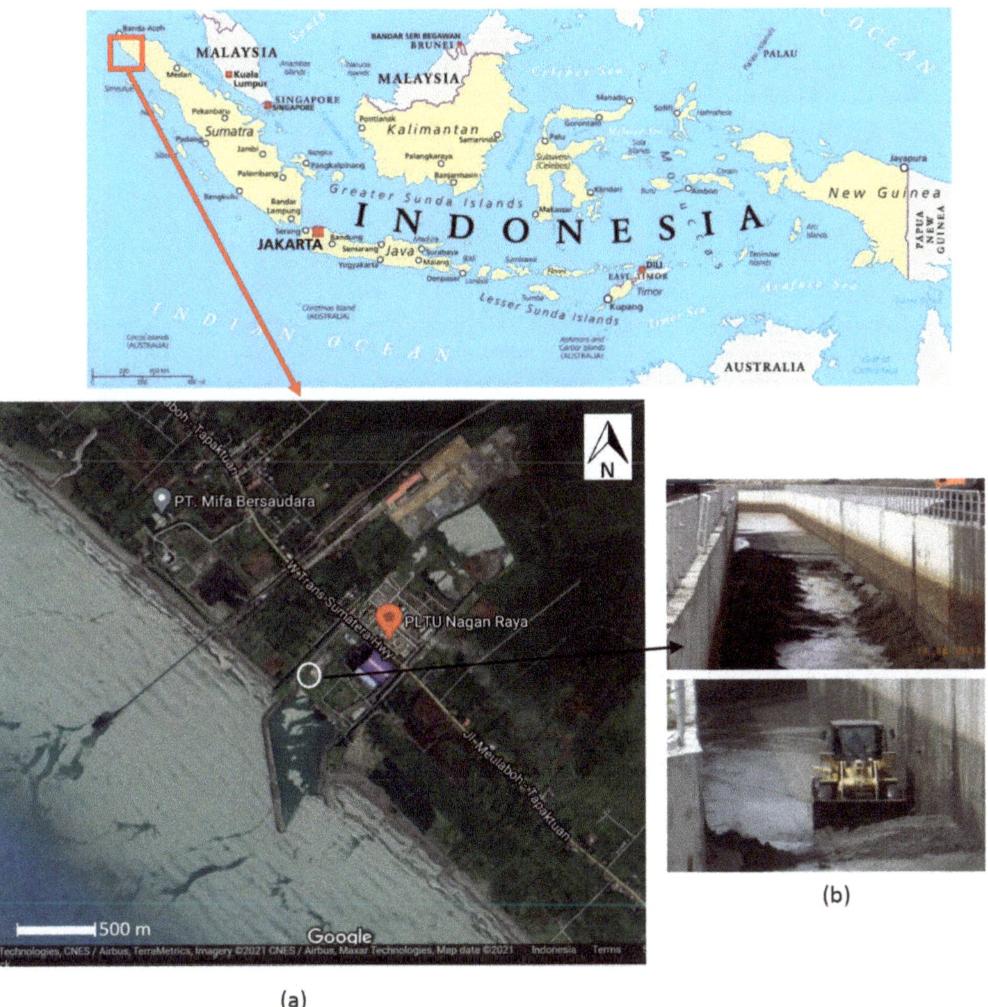

Figure 1. (**a**) Location of Nagan Raya Power Plant [18] (**b**) Sedimentation in water-intake channel.

3. Materials and Methods

3.1. Field Data

Field measurements of various parameters such as bathymetry, wave height, current speed, water level variation and sedimentation were executed in the area around Nagan Raya port. This field data was used in this study in order to achieve better knowledge of sedimentation patterns and to investigate the effects of these parameters on sediment supply in the area of interest. Previous sediment survey data were studied together with hydro-oceanographic data to develop a practical investigative plan for new sampling inside and outside of the port area. In this study, field measurements were taken in October 2018. The area of the hydrographic survey is shown in Figure 2a.

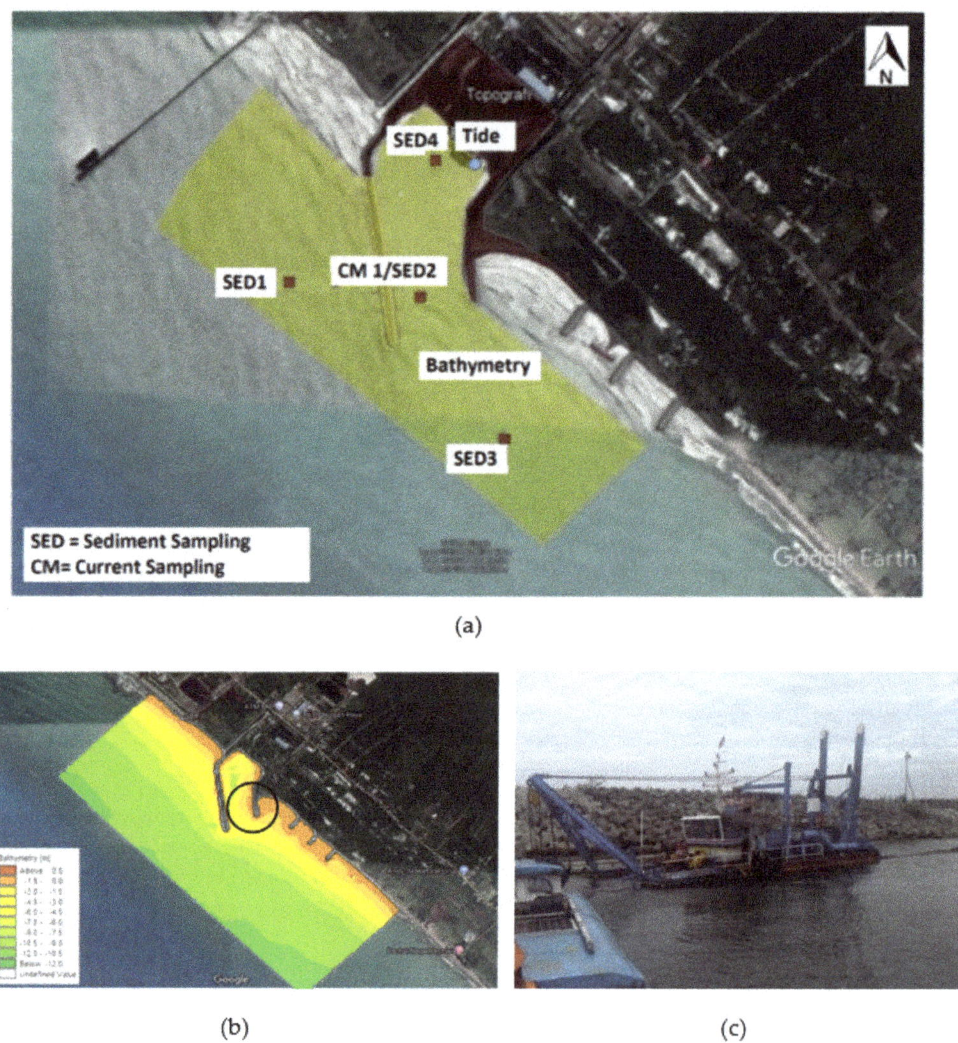

Figure 2. (**a**) Hydrographic Survey Area. (**b**) Bathymetric map of Nagan Raya port from 2018. (**c**) Dredging activity at port area.

The bathymetric map, as shown in Figure 2b, was measured on 24–25 October 2018, with the width of the area outside of the basin under measurement being approximately 1700 m. The depth displayed on the bathymetric map below refers to the Lower Water Spring (LWS). The area with the highest sedimentation is shown by the circle in Figure 2b.

In order to maintain the functioning of the port, routine dredging operations were performed at the port entrance on a daily basis, as seen in Figure 2c. During daily dredging operations, more than 50,000 m^3/month of sediments were removed from the navigation channel. The port entrance was dredged to a depth of 6 m. However, this dredging operation depended on the climate conditions of the waves. This was the case because of the location of Nagan Raya port, which is located directly facing the Indian Ocean with strong winds and high wave conditions. When the wave conditions are extreme, dredging activities at the port inevitably stop until environmental conditions return to normal. However, the sedimentation process continually appears in the port. This makes dredging maintenance inefficient and expensive.

Wind data for the years 2004–2018 were obtained from wind measurement stations of the Meteorology, Climatology and Geophysics Department [19] in 2004–2018. The wind speed and direction over Nagan Raya coast are shown in Figure 3b below. The average wind speed was 4.28 (m/s). The winds most frequently came from the south (15.6%), south-west (27.7%) and west (18.8%).

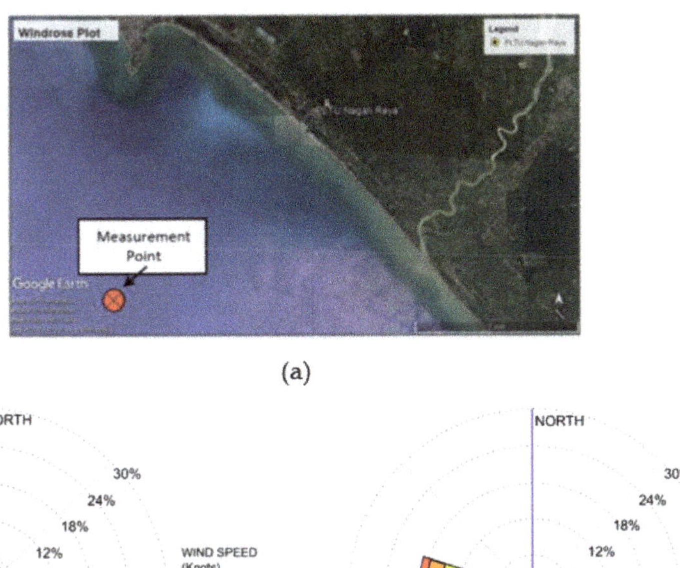

Figure 3. (**a**) Location of measurement point of wind data, (**b**) Wind Rose, and (**c**) Wave Rose in Nagan Raya port.

The wave data for the area around the Nagan Raya port were analyzed from wind climate data representing 10 year periods from 2004 to 2018. The dominant direction of the incoming wave was from the southwest. The significant wave height, denoted as *Hs*, was 2 m, and the typical wave period, represented by *T*, was 6.3 s. The probability distribution of the wave data is shown in Figure 3c.

In this study, current data were used for the calibration and validation of the numerical model. The current data used in this study were obtained at a depth point of 6–7 m, and represented an area located near the inlet of the port, as shown in Figure 2a. The current measurement was surveyed at 1-h intervals using the Valeport 106 current meter. The maximum observed current speed in this station was around 0.20 m/s and occurred at a depth of 0.2D or 3.5 m below the water surface. The maximum current speed was found for the current flowing from the southwest and the south, as shown in Figure 4.

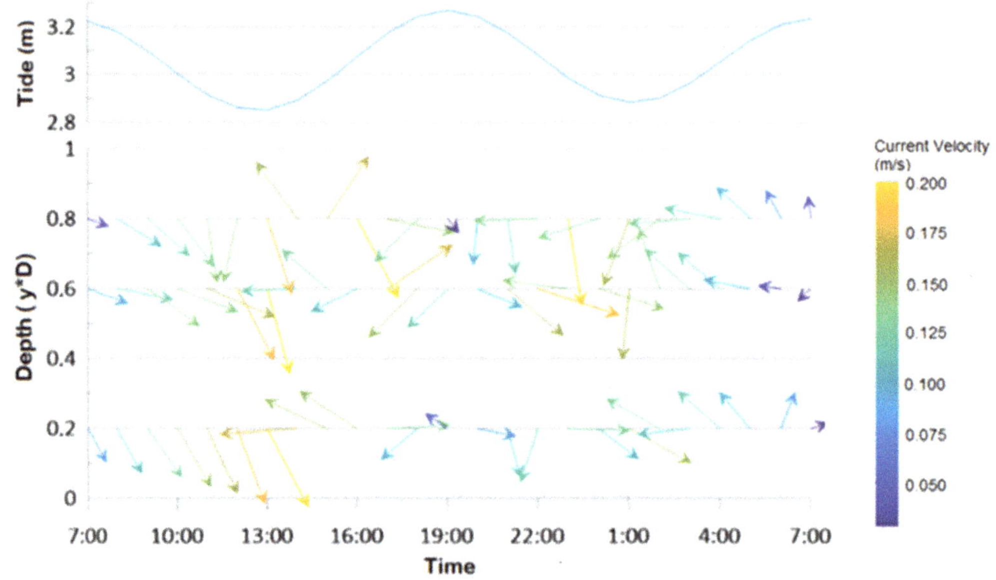

Figure 4. Current velocity at Nagan Raya Port.

Tidal measurement was conducted during the bathymetric measurement process as a means of correcting the sounding readings. From the tidal measurements, the typical tidal range in this area was found to be about 0.7 m. The tidal type of the Nagan Raya port area was mixed tide, with a prevailing semidiurnal pattern, and had a Formzahl value of $F = 1.05$. Table 1 provides a list of the tidal constituents calculated around the port of Nagan Raya.

Table 1. Tidal harmonic constants.

A (cm)	S_0	M_2	S_2	N_2	K_1	0_1	M_4	MS_4	K_2	P_1
	306	7	6	1	9	5	0	0	2	3
g°	0	153	249	170	285	217	306	22	249	285

Figure 2a shows the location in which bed sediment samples were taken around port basin. The results of the sediment laboratory test are shown in Table 2 below. Table 2 indicates that the fine sedimentary material was typically made up of sand. The data for the suspended sediment that were retrieved during the investigation conducted around port area shows typical ranges from about 40 to 150 mg/L.

Table 2. Sediment laboratory test.

No	Gravel	Sand	Silt	Clay	D50	Cu	Cc
1	0.00%	91.27%	8.73%	0.00%	0.168	2.641	0.998
2	0.00%	96.92%	3.08%	0.00%	0.197	2.374	0.904
3	0.00%	76.30%	23.70%	0.00%	0.097	2.55	1.729
4	0.00%	79.98%	20.02%	0.00%	0.15	4.763	1.513

3.2. Shoreline Changes

Several steps were taken in finding a constructive sedimentation solution for Nagan Raya Port. These included an investigation of historical sedimentation data, which can provide a good conception of the coastal processes that occur and how they affect the study area and the locations that surround it. Historical satellite images were used to manually extract the shoreline locations from 2011 to 2017 using Google Earth, as shown in Figure 5 below. The aim of this historical study was to evaluate and simulate the effect of the breakwater structure in terms of coastline changes.

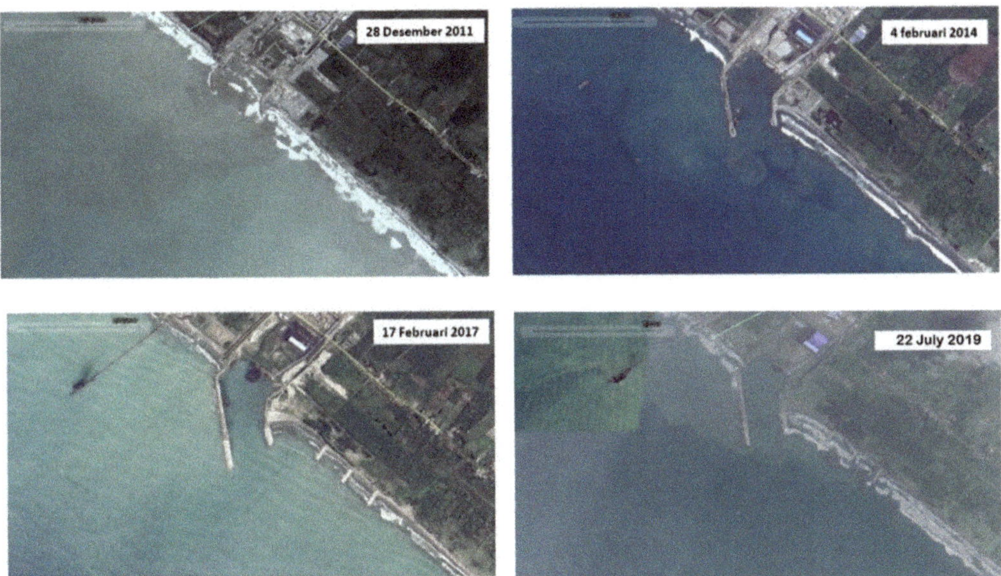

Figure 5. Historical coastline changes (2011–2019).

In this study, a new shoreline simulation model (ShorelineS), proposed by Roelvink et al. (2018, 2020) [20,21], was used for predicting the evolution of the coastline over 7 year periods. The basic equation for this ShorelineS model is based on sediment conservation, details about which are provided in Roelvink et al. (2020) [21].

For the setup of the ShorelineS model, the initial shoreline in 2011 and the structure of the breakwater were applied as land boundaries. The total length of the coastline was 9.25 km. An initial grid size of 5 m and the CERC1 formula derived from USACE (1984) [22] were applied with a closure depth of 10 m. The total simulation time was 10 years. A mean wave direction of southwest, with +/− 15 degrees of variation, was applied, with an average wave height of 1.7 m and a peak wave period of 6.3 s. The results of the coastline change simulation are shown in Figure 6.

To obtain an idea of the effects of sediment transport on the shoreline due to waves, it was necessary to analyze the rate of sediment transport. The flow rate of sedimentation along a coast depends on the angle of incidence of the waves, as well as the duration and energy of the waves. Thus, large waves will carry more material per unit of time when driven by small waves. Sediment transport analysis was carried out to determine the flow rate of sedimentation along the coast using the CERC method. The formula of the CERC method [22] can be expressed as follows:

$$Qs = k \times P1^{n} \quad (1)$$

$$P1 = \frac{\rho g}{8} \times H_{b2} \times C_{b} \times \cos ab \times \sin ab \quad (2)$$

where Qs is sediment transport along the coast (m^3/day). $P1$ is a component of the wave energy flux at break (Nm/s/m). ρ is a sea water mass of 1025 kg/m^3. H_b is height of the breaking wave (m). C_b is the broken wave propagation (m/s). *ab* represents the angles of the breaking waves. *n* is a constant of 1.29. *g* is a gravity acceleration of 9.81 m/s^2. The results indicate that the rate of longshore sediment transport was 58,576 m^3/month.

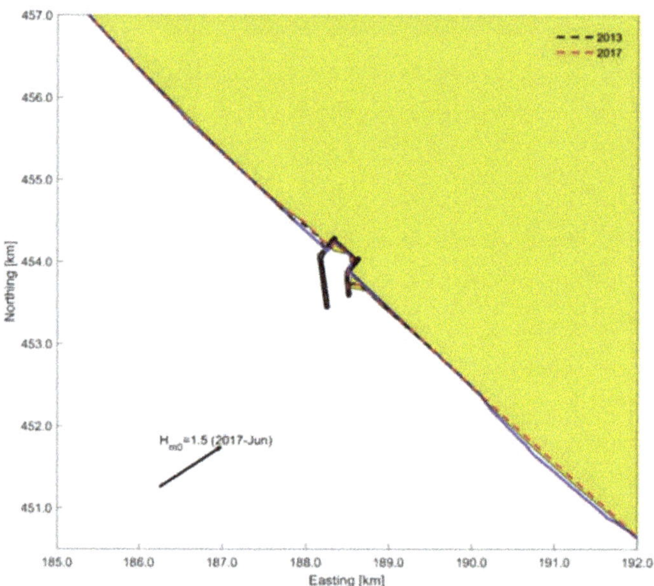

Figure 6. Coastline change simulation result.

Calculation of the sedimentation flow rate was also carried out using sounding bathymetric data by overlaying bathymetric data at different times. This analysis used bathymetric measurement data collected in July 2018 (Figure 7a) and August 2018 (Figure 7b). Through the calculation of the overlay volume, it was found that the total net volume was 40,194 m3 (Figure 7c). Thus, it can be seen that the sediment transport rate in July 2018–August 2014 was 40,194 m^3/month, as shown in Figure 7c. The calculation result obtained using the CERC formula was greater than the overlaid bathymetry result, which was the case because the incoming wave used in the analysis was that of only one direction, namely from the southwest direction.

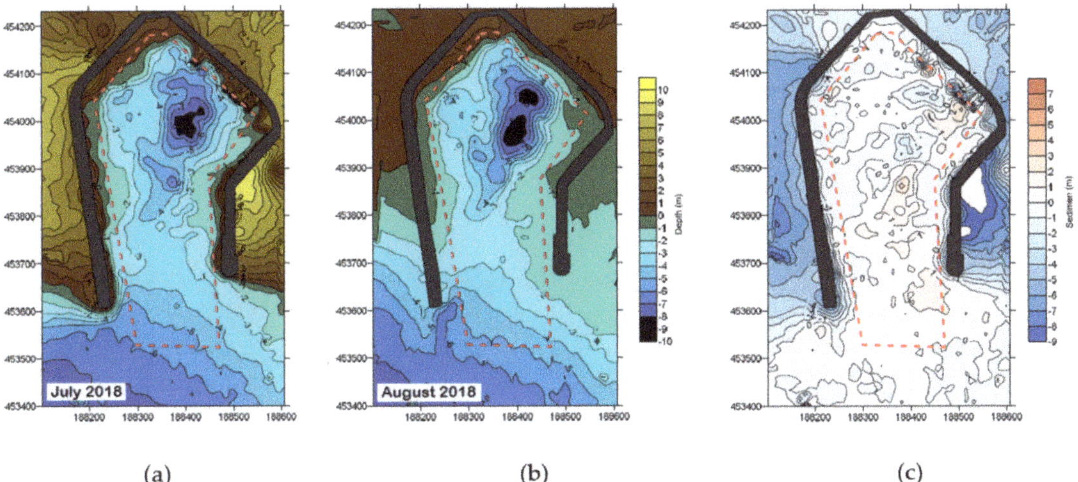

(a) (b) (c)

Figure 7. Bathymetric measurement map: (**a**) July 2018, (**b**) August 2018. (**c**) Overlaid bathymetry.

4. Delft 3D Model

In this study, the Delft3D process-based modeling system was used to simulate the sedimentation pattern around the Nagan Raya port area. The Delft3D model is a useful tool in the development of effective sedimentation solutions. In addition, the Delft3D software package has several different modules. This model capable of simulating aspects of waves, currents, sediment transport, morphological developments and water quality in coastal and ocean areas [23]. The Delft3D-Flow model [24–26], through the use of the third-generation Simulating Waves Nearshore Model (SWAN) [27–30], and the Delft3D-Wave module are both used to simulate the transformation of short waves. Details regarding the equations used in, and the practical use of, the Delft3D modeling system can be found in the DELFT3D user manuals [24,27]. The Delft3D model, and the Delft3D-Flow [24,25] and Delft3D-Wave modules [27,28], were applied in this study.

The grid system for domain computation in the Delft3D system is shown in Figure 8. The application of the bathymetry model and the measurement of land boundaries were mainly based on field measurement data obtained during the study. In this study, curvilinear grids were used in the computational domain with different resolutions. The grid sizes resolution varied from 10 to 500 m, with the coarser grid at the offshore boundary (100–500 m) and a finer resolution inside the port (10–50 m). In total, three open boundaries were used in the domain: the water level in the southern area and the Neumann boundary in the eastern and western areas, as shown in Figure 9.

After the initial setup of the model, the next step was to calibrate and validate the results of model with observed data. This stage was useful for evaluating the performance of the model by adjusting the model-free parameters, and thus, establishing whether the existing modeling data were accurate enough to be used at a later stage. In the flow model, model parameters such as the bed roughness coefficient and the eddy viscosity should be optimally adjusted to produce model results that are as similar as possible to the actual measured values. In addition, improvements in the size of the computational grid can also affect the accuracy of the results of the model [31–33]

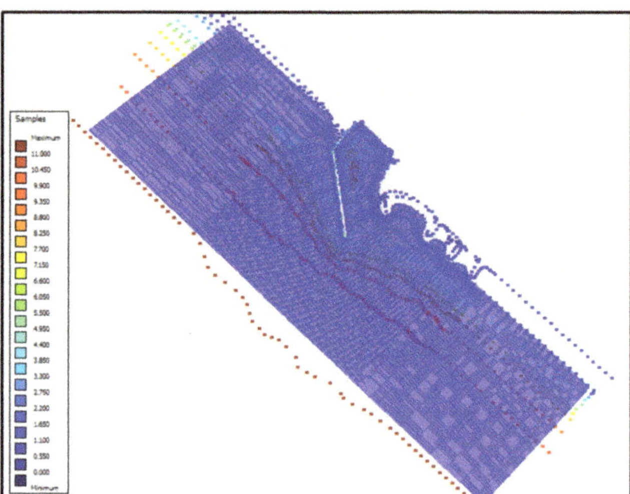

Figure 8. Curvilinear grids and mesh for Nagan Raya port.

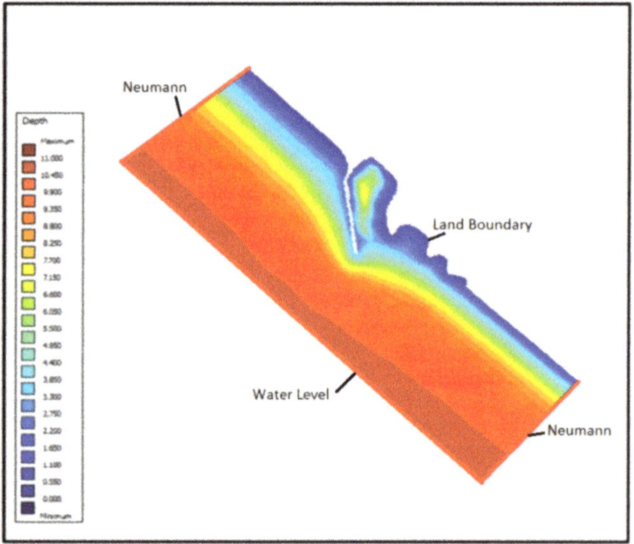

Figure 9. Model domain and boundary conditions.

In this study, water level gauge measurement was used for calibration of the model, as presented in Figure 10. The calibrated water level data showed a close relation and a good agreement between the observed and simulated water level, as presented in Figure 10. After calibration, the model was validated by the comparing simulated and observed current velocity data, as shown in Figure 11. The simulation results for the current velocity generally had the same pattern and were quite close to the values obtained via measurement. The RMSE (root mean square error) statistical parameter between the simulated and observed current velocity was 0.0003. Thus, the model performance was considered to be good enough and it was determined that the model could be used for the purposes of this study.

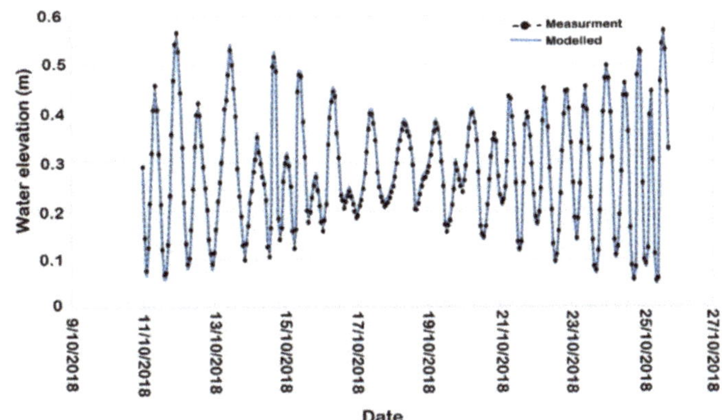

Figure 10. Comparison between modelled and measured water elevation.

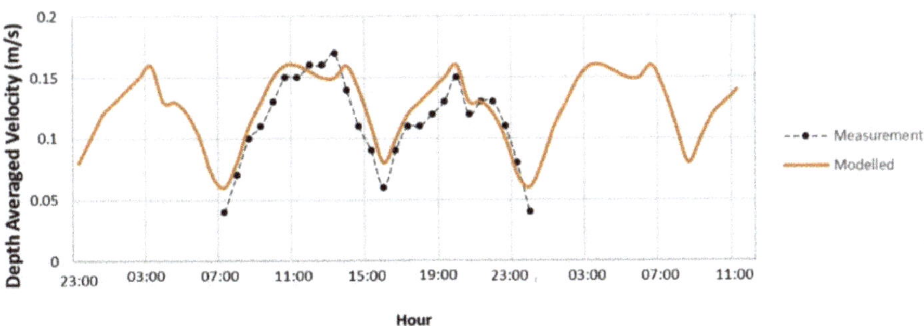

Figure 11. Comparison of depth-averaged velocity between the modelled (solid line) and the observed (dot line) data.

The final input parameters used for the Delft3D model simulation are shown in Tables 3 and 4 below. In this model, modeling was conducted for a 1-month simulation period.

Table 3. Hydrodynamic parameters.

Hydrodynamic Parameters	Value
Model configuration	2DH (depth-averaged)
Number of grid elements	8960
Meteorological forcing	Astronomical tidal forcing, wind data
Time step	60 s
Gravity	9.81 m/s^2
Water density	1025 kg/m^3
Roughness (Chezy coefficient)	7
Horizontal eddy viscosity	1 m^2/s
Horizontal eddy diffusivity	1 m^2/s

Table 4. Morphological parameters.

Morphological Parameters	Value
Specific density	2650 kg/m^3
Dry bed density	1600 kg/m^3
Median sediment diameter (D_{50})	200 μm
Initial sediment layer thickness at bed	5 m
Threshold sediment depth thickness	0.5 m

5. Results and Discussion

The results obtained using the sediment transport model are shown in Figure 12. Based on the sedimentation modelling results for the port basin, as shown in Figure 12, it was found that around the mouth of the basin area, the sedimentation occurred at levels as high as 0.8–1 m. This is supported by the bathymetric measurement data in this area; around the coastline at the near side of the breakwater, there was sedimentation as high as 1.4–1.6 m. In addition, it was found that the sedimentation rate at the harbor basin was around 38,346 m^3/month. This value was found to be slightly underestimated based on the averaged dredged volumes that were obtained via bathymetric measurement. The choice of different parameters used in the sediment transport model might have been the cause of this result. Furthermore, it is possible that errors occurred during the bathymetric measurement process.

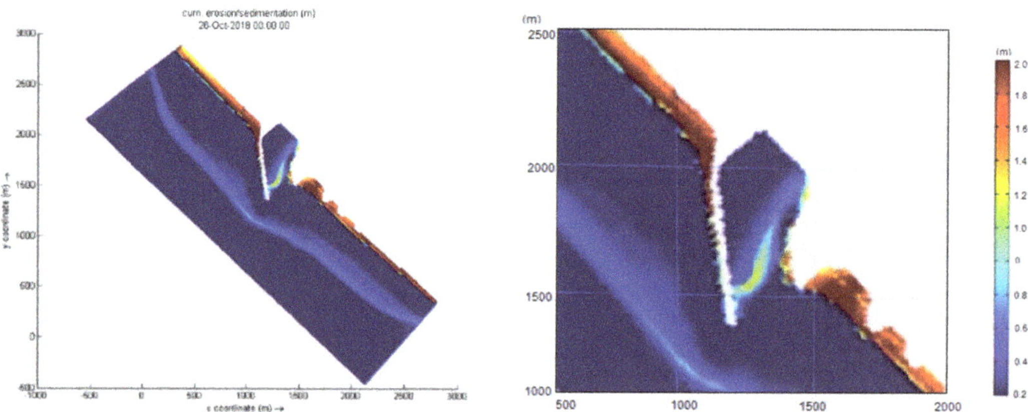

Figure 12. Erosion and sedimentation model after 1 month of simulation.

In addition to the above modeling results, the dredging and dumping activities could be simulated in Delft3D software package. To evaluate the impacts of sedimentation on the adjacent port area, 1-month morphology change simulations were conducted for the dredging scenario of the navigation channel using the dredging and dumping utility in Delft3D [24]. During simulation, the characteristics of the dredging and dumping activities that occurred were saved in a dredge and dump file. A detailed description of the dredging and dumping features in this model can be found in the Delft3D-Flow user manual [24].

In this study, two dredging activity scenarios were applied using the Dredge and Dump feature of Delft3D. The volume of sediment was automatically removed from one area in the bathymetry model and dumped into the dump area. For the first scenario, the dredged material was dumped in Location 1 in the offshore area. For the second scenario, the dredged sediment was dumped in Location 2 in the near side of the breakwater, as shown in Figure 13. In this modeling system, it was assumed that the sedimentation was dredged and maintained at a 6 m depth as a threshold water depth. This approach was adapted for coal barges that often pass through the basin area of the Nagan Raya port. The results of the dredging and dumping modeling process are presented in Figure 14 below:

Figure 13. Dredging and dumping location.

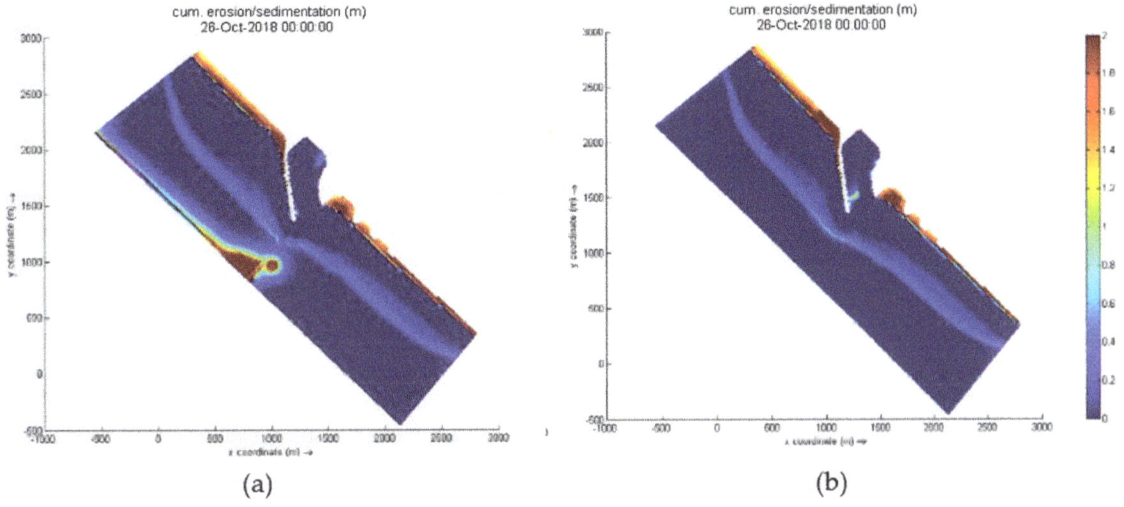

Figure 14. Dredging and dumping modeling: (**a**) scenario 1 and (**b**) scenario 2.

For scenario 1 (first), the dumping point was located at a depth of 11 m and the distance from the coastline was about 1.5 km. The dredged material was sucked up by the dredger and channeled through the pipe to the shelter barge. The fully loaded barge was pulled by a tugboat, at a speed of 3 knots, to the dredge disposal area (scenario 1), as shown in Figure 14a. Based on the modeling results for scenario 1, the dredged material that had been disposed of in the dumping area did not return to the mouth of the basin area.

In contrast, for scenario 2, after the removal of the dredge material to Location 2 in the north part of the breakwater side, there was still sedimentation in the mouth area of the basin, as shown in Figure 14b. Figure 14b shows the potential locations of the dredged material that was dumped back into the mouth of the basin area. Based on the above results, Area 1, located in the offshore area, was proposed as the best dumping area location. In addition to complying with government regulations, the proposed disposal area must also reach a further 20 miles seaward from the shoreline, or a minimum depth of 20 m.

6. Conclusions

The port of 2 × 110 MW Nagan Raya Coal Fired Steam Power Plant is a facility in Aceh Province, Indonesia, which was constructed by the State Electricity Company. During its operation, which began in 2013, the port has dealt with large amounts of sedimentation within the port and ship entrances. The existing breakwater construction has not been maximally effective in protecting the port basin from sedimentation. In addition, the breakwater structure and some groins have also been extended to minimize this problem. However, an enormous amount of material sediment is still dredged from the mouth of navigation channels in order to maintain port operations.

Evaluation of the wind data showed that the dominant wind direction was from south to west. Based on the analysis of the wave data, it is known that the dominant wave direction was from the south to the west. Therefore, wave-induced currents in the surf zone were from south to north. Based on the analysis of longshore sediment transport, the supply of sediment to Nagan Raya port was estimated to be around 40,000–60,000 m^3 per year. The large estimated amount of sediment transport along the coast of the port was based on wave direction from the south to the west.

Results obtained from the sediment model showed that sedimentation of up to 1 m was captured in areas of the ship entrance of Nagan Raya port. The sand bypassing system is one of the sedimentation management solutions that was proposed in this study. The dredged sediment material located around the ship entrance was dumped in a dumping area, located in the middle of the sea, at a depth of 11 m, with a distance of 1.5 km from the shoreline. To obtain a greater maximum result, the material disposal distance should be further away, at a depth of at least 20 m or a distance of 20 miles from the coastline. The proposed solution could reduce operating costs since the dredging operations will be localized to one specific area.

Author Contributions: Conceptualization, M.Z.; methodology, M.Z. and K.S.; validation, S.S.; writing—original draft preparation, M.Z. and S.S.; writing—review and editing, M.Z.; visualization, M.Z. and K.S. All authors have read and agreed to the published version of the manuscript.

Funding: The authors gratefully acknowledge the financial support received from the Institut Teknologi Sepuluh Nopember for this work, under project scheme of the Publication Writing and IPR Incentive Program (PPHKI).

Conflicts of Interest: The authors declare no conflict of interest and the funders had no role in the design of the study; in the collection, analyses, or interpretation of data; in the writing of the manuscript, or in the decision to publish the results.

References

1. PT. *Horas Bangun Persada. Kajian Laju Sedimentasi Pada Kolam Pelabuhan PLTU 2 × 110 MW Nagan Raya*; PT PLN (Persero): Jakarta, Indonesia, 2013.
2. Zhong-hua, T.; Hai-cheng, L.; Feng, G. Experimental Research on Reduction Measures of Sediment Deposition of the Power Plant Port under the Long Period Wave. *Procedia Eng.* **2015**, *116*, 229–236. [CrossRef]
3. Tianjin Research Institute. *Nagan Raya 2 × 110 MW Coal-Fired Power Plant, Indonesia*; Wave-Sediment Physical Model Test Report; Water Transport Engineering of Transport Ministry: Tianjin, China, 2015.
4. Parchure, T.M.; Teeter, A.M. *Potential Methods of Reducing Shoaling in Harbors and Navigation Channels*; HETN-XIV-6; Engineer Research and Development Center: Vicksburg, MS, USA, 2002.
5. Salman, A.; Lombardo, S.; Doody, P. *Living with Coastal Erosion in Europe: Sediment and Space for Sustainability*. 2004. Report Commissioned by the European Commission. Available online: http://www.eurosion.org/reports-online/part4.pdf (accessed on 13 July 2021).
6. Gardner, W.D. Sediment trap dynamics and calibration: A laboratory evaluation. *J. Mar. Res.* **1980**, *38*, 17–39.
7. Botwe, B.O.; Abril, J.M.; Schirone, A.; Barsanti, M.; Delbono, I.; Delfanti, R.; Nyarko, E.; Lens, P.N.L. Settling fluxes and sediment accumulation rates by the combined use of sediment traps and sediment cores in Tema Harbour (Ghana). *Sci. Total Environ.* **2017**, *609*, 1114–1125. [CrossRef] [PubMed]

8. Hofland, B.; Christiansen, H.; Crowder, R.A.; Kirby, R.; van Leeuwen, C.W.; Winterwerp, J.C. The current deflecting wall in an estuarine harbor. In Proceedings of the XXIV IAHR Congress, Beijing, China, 16–21 September 2001.
9. Van Kessel, T.; Cornelisse, J.M. *Physical Scale Model Zeeschelde*; Delft Hydraulics, Report Z2516; WL Delft Hydraulics: Delft, The Netherlands, 2003.
10. Juez, C.; Bühlmann, I.; Maechler, G.; Schleiss, A.J.; Franca, M.J. Transport of suspended sediments under the influence of bank macro-roughness. *Earth Surf. Process. Landf.* **2018**, *43*, 271–284. [CrossRef]
11. Weisman, R.N.; Lennon, G.P.; Clausner, J.E. *A Guide to the Planning and Hydraulic Design of Fluidizer Systems for Sand Management in the Coastal Environment*; Dredging Research Program; US Army Corps of Engineers: Washington, DC, USA, 1996; Technical Report No. DRP-96-3.
12. Jones, C.P.; Mehta, A.J. Inlet sand bypassing systems in Florida. *Shore Beach* **1980**, *36*, 27–30.
13. Ware, D. *Tweed River Entrance Sand Bypass Project*; Case Study for Coast Adapt; National Climate Change Adaptation Research Facility: Gold Coast, Australia, 2016.
14. Mendes, D.S.; Fortunato, A.B.; Pires-Silva, A.A. Assessment of three dredging plans for a wave-dominated inlet. *Proc. Inst. Civil Eng. Mar. Eng.* **2016**, *169*, 64–75. [CrossRef]
15. Reyes-Merlo, M.A.; Ortega-Sánchez, M.; Díez-Minguito, M.; Losada, M.A. Efficient dredging strategy in a tidal inlet based on an energetic approach. *Ocean Coast. Manag.* **2017**, *146*, 157–169. [CrossRef]
16. Shaeri, S.; Tomlinson, R.; Etemad-Shahidi, A.; Strauss, D. Numerical modelling to assess maintenance strategy management options for a small tidal inlet. *Estuar. Coast. Shelf Sci.* **2017**, *187*, 273–292. [CrossRef]
17. Fernández-Fernández, S.; Ferreira, C.C.; Silva, P.A.; Baptista, P.; Romão, S.; Fontán-Bouzas, Á.; Abreu, T.; Bertin, X. Assessment of Dredging Scenarios for a Tidal Inlet in a High-Energy Coast. *J. Mar. Sci. Eng.* **2019**, *7*, 395. [CrossRef]
18. Google Earth. Map Showing Location of Nagan Raya Power Plant. Available online: https://earth.google.com (accessed on 6 August 2021).
19. BMKG. Climatology Data on Nagan Raya. Available online: https://dataonline.bmkg.go.id (accessed on 15 December 2019).
20. Roelvink, D.; Huisman, B.; Elghandour, A. Efficient modelling of complex coastal evolution at monthly to century time scales. In Proceedings of the Sixth International Conference on Estuaries and Coasts (ICEC-2018), Caen, France, 20–23 August 2018.
21. Roelvink, D.; Huisman, B.; Elghandour, A.; Ghonim, M.; Reyns, J. Efficient Modeling of Complex Sandy Coastal Evolution at Monthly to Century Time Scales. *Front. Mar. Sci.* **2020**, *7*, 535. [CrossRef]
22. USACE. *Shore Protection Manual*; US Army Coastal Engineering Research Center: Washington, DC, USA, 1984; Volume 1.
23. Roelvink, J.A.; Van Banning, G.K.F.M. Design and development of Delft3D and application to coastal morphodynamics. In Proceedings of the Hydroinformatics '94 Conference, Delft, The Netherlands, 19–23 September 1994.
24. Deltares. *Delft3D-Flow User Manual*; Version 3.14, Revision 12556; Deltares: Delft, The Netherlands, 2010.
25. Lesser, G.R.; Roelvink, J.A.; van Kester, J.A.T.M.; Stelling, G.S. Development and validation of a three-dimensional morphological model. *Coast. Eng.* **2014**, *51*, 883–915. [CrossRef]
26. Lesser, G.R. An Approach to Medium-Term Coastal Morphological DeltaresModeling. Ph.D. Thesis, UNESCO-IHE & Delft University of Technology, Delft, The Netherlands, 2009.
27. Deltares. *Delft3D-Wave User Manual*; Version 3.04, Revision 11114; Deltares: Delft, The Netherlands, 2010.
28. Booij, N.; Ris, R.C.; Holthuijsen, L.H. A third-generation wave model for coastal regions. Part 1: Model description and validation. *J. Geophys. Res.* **1999**, *104*, 7649–7666. [CrossRef]
29. Holthuisjen, L.H. *Waves in Oceanic and Coastal Waters*; Cambridge University Pres: Cambridge, UK, 2007.
30. Van der Westhuysen, A.J. Modelling nearshore wave processes. In Proceedings of the ECMWF Workshop on Ocean Waves, Reading, UK, 25–27 June 2012.
31. Pu, J.H. Conceptual Hydrodynamic-Thermal Mapping Modelling for Coral Reefs at South Singapore Sea. *Appl. Ocean. Res.* **2016**, *55*, 59–65. [CrossRef]
32. Jahid Hasan, G.M.; van Maren, D.S.; Cheong, H.F. Improving hydrodynamic modeling of an estuary in a mixed tidal regime by grid refining and aligning. *Ocean. Dyn.* **2012**, *62*, 395–409. [CrossRef]
33. Pu, J.; Huang, Y.; Shao, S.; Hussain, K. Three-Gorges Dam Fine Sediment Pollutant Transport: Turbulence SPH Model Simulation of Multi-Fluid Flows. *J. Appl. Fluid Mech* **2016**, *9*, 1–10. [CrossRef]

Article

Experimental Characterization of the Flow Field around Oblong Bridge Piers

Ana Margarida Bento [1,2,3,*,†], Teresa Viseu [1], João Pedro Pêgo [2,3] and Lúcia Couto [1]

1 Water Resources and Hydraulic Structures Division, Hydraulics and Environment Department, National Laboratory of Civil Engineering, 1700-066 Lisboa, Portugal; tviseu@lnec.pt (T.V.); luciacoutoribeiro@gmail.com (L.C.)
2 Hydraulics, Water Resources and Environmental Division, Civil Engineering Department, Faculty of Engineering of the University of Porto, 4400-465 Porto, Portugal; jppego@fe.up.pt
3 Interdisciplinary Centre of Marine and Environmental Research of the University of Porto, 4450-208 Matosinhos, Portugal
* Correspondence: ana.bento@fe.up.pt
† Current address: University of Minho, ISISE, School of Engineering, Civil Engineering Department, 4800-058 Guimarães, Portugal.

Abstract: The prediction of scour evolution at bridge foundations is of utmost importance for engineering design and infrastructures' safety. The complexity of the scouring inherent flow field is the result of separation and generation of multiple vortices and further magnified due to the dynamic interaction between the flow and the movable bed throughout the development of a scour hole. In experimental environments, the current approaches for scour characterization rely mainly on measurements of the evolution of movable beds rather than on flow field characterization. This paper investigates the turbulent flow field around oblong bridge pier models in a well-controlled laboratory environment, for understanding the mechanisms of flow responsible for current-induced scour. This study was based on an experimental campaign planned for velocity measurements of the flow around oblong bridge pier models, of different widths, carried out in a large-scale tilting flume. Measurements of stream-wise, cross-wise and vertical velocity distributions, as well as of the Reynolds shear stresses, were performed at both the flat and eroded bed stages of scouring development with a high-resolution acoustic velocimeter. The time-averaged values of velocity and shear stress are larger in the presence of a developed scour hole than in the corresponding flat bed configuration.

Keywords: bridge pier; flat and eroded bed; flow field; velocity profile measurements

Citation: Bento, A.M.; Viseu, T.; Pêgo, J.P.; Couto, L. Experimental Characterization of the Flow Field around Oblong Bridge Piers. *Fluids* **2021**, *6*, 370. https://doi.org/10.3390/fluids6110370

Academic Editor: Jaan H. Pu

Received: 7 September 2021
Accepted: 12 October 2021
Published: 20 October 2021

Publisher's Note: MDPI stays neutral with regard to jurisdictional claims in published maps and institutional affiliations.

1. Introduction

The presence of obstacles such as bridge foundations in rivers is often associated with local morphological changes as a result of altering the sediment–flow equilibrium at the vicinity of those hydraulic infrastructures [1,2]. The modification of local geomorphological bed characteristics, known as local scour, should be accounted for when analysing the safety of bridges due to the severe threat that the uncovering of the foundation can pose to its integrity. Local scour around piers is one of the major threats to bridge structures founded in riverbeds and is the result of the complex flow patterns around those foundations [3–5]. It is therefore important to be acquainted with the flow structure and the related scour mechanisms around bridge piers.

The formation of highly turbulent vortex systems leads to increased local velocities and bottom shear stress, which results in the formation of scour holes. The sediment entrainment and transport are then controlled by the turbulent flow field therein developed [6,7]. Once formed, the scour hole interferes significantly with the turbulent flow field and the hydrodynamic characteristics [8,9]. For instance, the bed shear stress and turbulence quantities are strengthened, which induces further sediment transport and scour development in the pier vicinity.

Besides the formulation of scour depth predictors, the understanding of complex turbulence patterns in the vicinity of bridge piers is also crucial in planning and designing phases of foundations' orientation, shape and geometry [6]. Several measuring techniques have been used before to characterize the flow field around piers. They began with the advent of propeller gauges, electromagnetic gauges, hot-film anemometers [10] and five-hole pitot spheres [11], and then developed to more sophisticated measuring equipment, such as Acoustic Doppler Velocimeter (ADV) [8,9,12,13], Laser Doppler Velocimetry (LDV) [14] and Particle Image Velocimetry (PIV) [15–17].

Ataie-Ashtiani and Aslani-Kordkandi [18] investigated the ADV measurements of turbulent flow field around single, side-by-side and tandem pile groups with and without a scour hole. Their results corroborated that the in-depth understanding of the flow and turbulence influenced by the scour hole is essential to improving the prediction of scour depth, see [19–21] and references therein.

The main flow processes that take place around a circular pier exposed to a steady current on a plane bed have been the subject of several flow field studies, both experimental and numerically, see [6], and references therein. They comprise the horseshoe vortices formed immediately upstream of the pier, the lee-wake vortices (usually in the form of vortex shedding), contracted streamlines at the side edges of the pier, and the downflow process due to the deceleration of the flow at the pier front. Dargahi [7] reported the result of several studies that focused on the visualization of the vertical structures around a bridge pier placed in an incoming fully turbulent flow at different stages of the scouring process. Graf and Yulistiyanto [22] measured the mean velocity components and turbulent Reynolds stresses using an ADV profiler in several planes around a circular pier for equilibrium scour conditions. By contrast, Roulund et al. [23] performed experimental and numerically visualizations of the flow in the scour hole during the evolution of the scour process and measured flow field and bed shear stresses for circular bridge piers on a flat bed. Dey and Raikar [8] have studied experimentally the structure of the turbulent horseshoe vortex flow within the developing scour holes at cylindrical piers by measuring instantaneous 3D velocities using an ADV. Notwithstanding, the flow in the rear part of the pier was not studied; they concluded that the flow and turbulence intensities in the horseshoe vortex remain rather similar during the development of the scour hole. While these flow structures are well-established for circular pier shape, few works have been devoted to the flow field study around bridge foundations of other shapes and geometries.

Recent studies have been focused on the characteristics of turbulence causing scour around circular, tandem, staggered arrangements and oblong piers [24,25], which did not account for the presence of scour holes. In Vijayasree et al. [24], the turbulent kinetic energy, turbulence intensities and Reynolds shear stresses were weaker for oblong piers, suggesting that this shape is a better alternative for the construction of bridge piers. The experimental work conducted by Pasupuleti et al. [25] was performed under clear water as the majority of previous research, which does not totally reproduce the field conditions. Therefore, the relatively limited number of flow field studies induced by the presence of an oblong bridge pier, most common shape of Portuguese bridge foundations of the 19th and 20th centuries, when compared with circular piers, motivated this investigation.

Therefore, the current study aims to identify and analyse the physics of the flow disturbed by the presence of a developed scour hole around oblong bridge piers. For that purpose, flow and turbulence characteristics are analysed, on the basis of an experimental campaign, concerning the change in pier effective width, approach hydraulic conditions and bed morphology stages, at the initial flat and the final eroded bed stages. Six fixed bed experiments, involving models of oblong bridge piers, were carried out in a recirculating tilting flume located in the experimental facilities of the National Laboratory of Civil Engineering (LNEC), in Lisbon, Portugal. Based on velocity measurements, the flow fields at the flume mid-plane surface were analysed, in terms of the time-averaged velocities and

Reynolds shear stresses. Furthermore, space-averaged quantities were also obtained to clarify the influence of scour development on the flow and turbulence characteristics.

Following this introduction, the flow field experiments are characterized in Section 2. The description of the experimental facility and bridge pier models, the devices, measured variables and instrumentation, experimental procedure and conditions are given in Section 2. A particular description of the approach used for the characterization of the turbulent flow field is included in Section 3. The results and their discussion are presented in Section 4. Finally, the main conclusions of this study are provided in Section 5.

2. Flow Field Experiments

2.1. Experimental Facility and Bridge Pier Models

The experiments were undertaken in the tilting flume of LNEC (Figure 1). The tilting flume, herein termed CIV, is 40.7 m long, 2.0 m wide and 1.0 m deep, with glass wall sides and a concrete base.

(a) Upstream working section (b) View from inside. (c) View from outside.

Figure 1. Experimental facility and bridge pier models.

Figure 2 presents the schematic setup of the flume. Two working sections were created within the base of the tilting flume, each being approximately 5.0 m long and 0.4 m deep. These sections were filled with uniform quartz sand characterized by a median diameter, D_{50}, of 0.86 mm, a density, ρ_s, of 2650 kgm^{-3} and a standard deviation of sediment particle sizes, σ_D, of 1.28. The working sections, herein also referred to as "sand recess boxes", are preceded by 2.0 m long accelerating ramps and 7.0 m long fixed bed reaches, and are followed by 3.0 m long fixed bed reaches. The accelerating ramps ("Acc. ramp" in Figure 2), placed at flume inlet and at the stilling recess boxes, conduce to uniform flow distributions in the two working sections. The stilling recess area between the two working sections was conceived to restore the flow regime for the downstream working section, and to store the sand bed material transported by the flowing water from the upstream working section during local scouring experiments. In the upstream boundary of the working sections, fine gravel mattresses, properly levelled by the respective concrete base, allow for smooth and progressive transitions, as well as ensure a turbulent flow development.

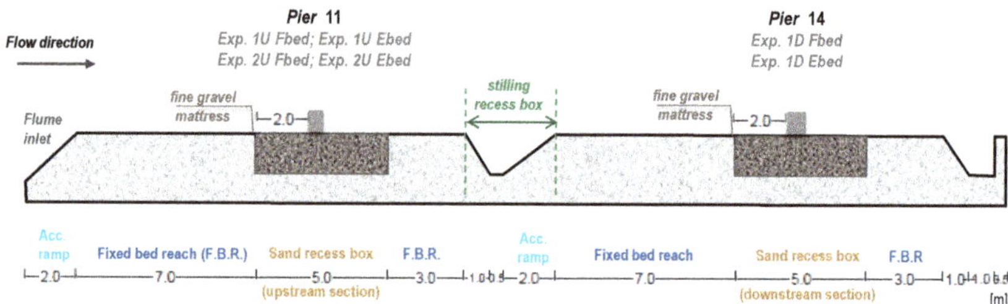

Figure 2. Schematic layout of the experimental setup (enlarged vertical scale).

Two oblong bridge pier models, *Pier* 11 and *Pier* 14, were placed in the upstream and downstream working sections, respectively. The piers differ in width (W), length (L), and semi-cylindrical surface ratio (R), schematically identified in Figure 3. The dimensions of *Pier* 11 are W = 0.11 m, L = 0.433 m and R = 0.065 m. *Pier* 14 is slightly larger, with W = 0.14 m, L = 0.463 m, and R = 0.08 m. The bridge pier models were placed vertically on the flume mid-plane at $\approx$ 2 m from the working sections' upstream borders (Figure 2). The contraction and wall effects were negligible to the scouring process, since the associated flume "width-to-pier width" ratios (W/B, B being the flume width) were 5.5% and 7.0% (for *Pier* 11 and *Pier* 14, respectively), both lower than the limit value of 10% indicated by Chiew and Melville [26].

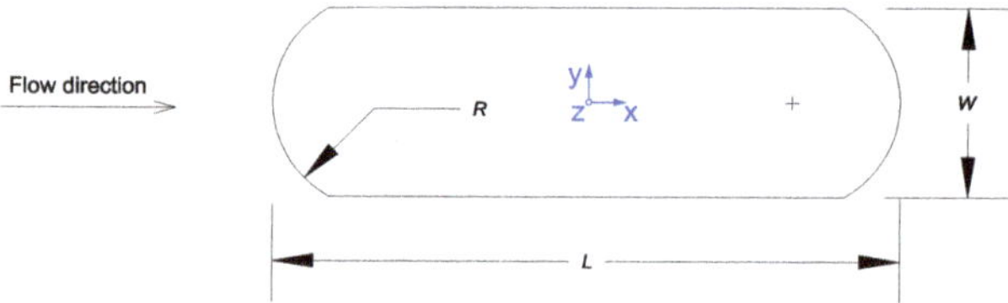

Figure 3. Geometry of the oblong bridge pier models.

The present study comprises the hydrodynamic characterization of six experiments divided into fixed flat bed and eroded bed stages of three local scouring experiments, namely *Exp.* 1*U*, *Exp.* 1*D* and *Exp.* 2*U*. The letters *U* and *D* denote the pier model, being *U* for *Pier* 11 and *D* for *Pier* 14. These local scouring experiments are characterized in detail in Bento [27,28]. For simplicity, a suffix was added to the local scouring experiments' name. The suffix *Fbed* stands for the fixed flat bed, whilst *Ebed* denotes the fixed eroded bed experiments. Therefore, the six flow field experiments were labelled as follows: *Exp.* 1*U Fbed* and *Exp.* 1*U Ebed* for *Exp.* 1*U*; *Exp.* 1*D Fbed* and *Exp.* 1*D Ebed* for *Exp.* 1*D*; and *Exp.* 2*U Fbed* and *Exp.* 2*U Ebed* for *Exp.* 2*U*.

2.2. Devices, Measured Variables and Instrumentation

Several flow variables were measured during the experiments, namely: (i) the flow discharge at the flume entrance; (ii) the flow depth within CIV and at the immediacy of the working sections; and (iii) the instantaneous three velocity components of the flow at the piers' vicinity. The following paragraphs describe the measuring process of each variable.

A variable frequency drive was used to control the rotational speed of the pump motor in order to obtain the desired flow discharge. The approach flow discharge was then monitored by an electromagnetic flowmeter with a rate accuracy of ±0.25%. At the flume entrance, stilling grids and a flow straightener were provided for damping the flow containing large scale eddies, leading to vortex free uniform flow in CIV. An adjustable sluice gate at the downstream end of the flume enables obtaining the desirable flow depth within the flume.

The approach flow depth was monitored using the acquired signals from acoustic probes, suitably positioned within the flume so as not to interfere with the development of the experiments. The signal of the three acoustic probes was acquired at a sampling frequency of 10 Hz, whereas the signal of the electromagnetic flowmeter was acquired at 25 Hz. Movable carriages were present on the rails of the flume to allow the access for controlling and measuring purposes of both the approach flow depth and the acquisition of the flow velocity components by the connection of the downlooking vectrino to its respective acquisition firmware (*Vectrino Plus*).

The instantaneous three-dimensional velocity measurements were acquired by using a high-resolution acoustic velocimeter, namely a downlooking vectrino (hereinafter designated as vectrino), from Nortek®, in two different moments: (i) at the beginning of the scouring process (flat bed stage), and (ii) at the final stage of the scouring experiments (eroded bed stage). The velocity flow field was measured throughout the working section around the pier models at relative elevations for the comparison of turbulence and flow characteristics due to the scouring process (before scour vs. after scour). An independent moving carriage system was developed to clamp the vectrino in order to move it along the different velocity measurement grid points to ease the measurement task, up to an accuracy of ±1 mm in the three directions (x, y and z). This issue will be considered in more detail in Section 3. The origin was considered at the pier centre, where x, y and z denote the longitudinal, transversal and vertical axes, respectively (see Figure 3).

2.3. Experimental Procedure

To facilitate the performance of the experiments on fixed beds, the surface layers of the bed material (Step 1—Figure 4a) were frozen by spraying uniformly a synthetic glue over the bed morphology stages (flat and eroded). This practice is frequently used in flow field studies under the assumption that the results are simpler and more explanatory than those from complex scenarios with a variety of acting processes, such as sediment burial and reappearance processes [29], and references therein.

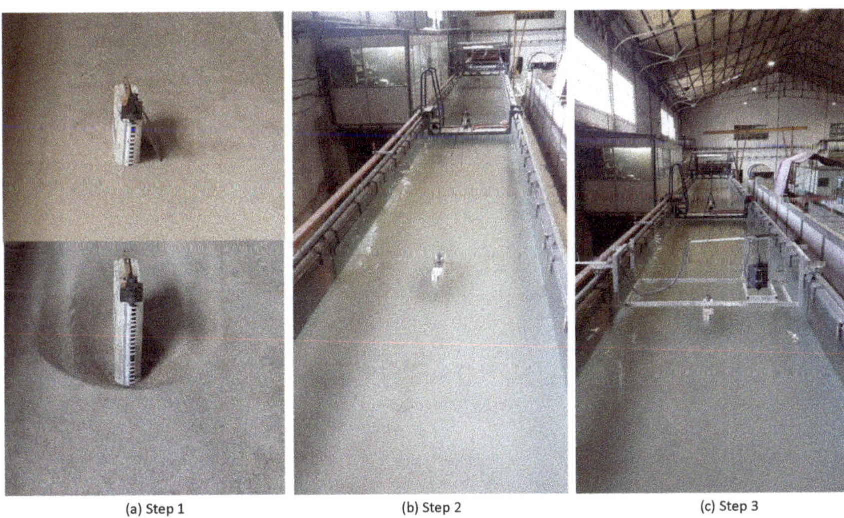

Figure 4. Setup for the flow velocity measurements.

Once the sand beds were properly stabilized, after approximately 72 h, the same hydraulic conditions of the corresponding local scouring experiments [27,28] were imposed, and the flume was filled up to the desired flow discharge and flow depth (conditions referred later). Since the vectrino uses the Doppler effect to measure current velocity by transmitting short pairs of acoustic waves, they must bounce off from particulate matter in the flow. As the laboratory water was quite clean, an artificial introduction of particles (commonly known as seeding particles) was deemed necessary, and thus a silica powder was added to the flow. An indicative aspect of the quality of the seeded flow is the parameter signal-to-noise ratio (SNR), provided by the *Vectrino Plus* interface; the higher the SNR, the better the seeding, and consequently, the more reliable the velocity measurements (Step 2—Figure 4b).

Five acoustic probe tips (one transmitter and four receivers, illustrated in Figure 4c) send and receive acoustic information in a fluid volume below the vectrino, referred to as the sampling volume. Since the instrument is inserted into the flow, the sampling volume must be sufficiently far away from the probe tip so as not to disturb the flow and the velocity measurements. Using acoustic waves, the vectrino takes real-time measurements (Step 3—Figure 4c). To evaluate quality of velocity measurements, one of the real-time outputs that vectrino provides is a correlation statistic. This is a value ranging from 0% to 100%; the closer the correlation is to 100%, the less noisy and more reliable a velocity measurement is. A minimum correlation of 70% is strongly recommended by several researchers [12], among others, and thus considered in the present study. In addition, a minimum signal-to-noise ratio (SNR) of 15 dB was designated for ensuring the reliability of the instantaneous velocity data [30,31].

Before conducting each experiment, several trials were undertaken to estimate the uncertainty in the measurement sampling duration. To tackle this, an assessment of the signal acquisition time's influence on the statistics of turbulent quantities was performed in Bento et al. [32]. Thus, signal acquisition times within the range of 120 to 300 s were considered in order to achieve time-independent velocity components, regarding the magnitude of turbulence at each measuring location. The vectrino was operated with an acoustic frequency of 10 MHz and a sampling rate of 200 Hz.

Besides the flow velocity characterization of the six fixed bed experiments by a comprehensive measuring system (resumed later), the flow depth was also continuously monitored for ensuring the predefined hydraulic conditions. After the completion of each fixed bed experiment, the surface layers of the sand beds were removed, the sand box re-levelled and once again fixed by the synthetic glue (to perform the corresponding flat bed experiment). A careful re-establishment of the hydraulic conditions was performed with a gradual filling of the flume.

2.4. Hydraulic Conditions and Time Duration

Table 1 presents the main hydraulic variables of the fixed bed experiments, namely the approach flow discharge (Q), flow depth (h), flow velocity (V), Froude number ($Fr = V/\sqrt{gh}$), flow Reynolds number ($Re = Vh/\nu$, being ν the kinematic viscosity of the fluid), and the parameter concerning the relation with the pier model: the pier Reynolds number ($Re_D = VW/\nu$). In Table 1, these parameters are given with a 95% of confidence interval; the estimated error corresponded to the sample error, expressed as $1.96\,\sigma/\sqrt{n_s}$ (being σ the standard deviation and n_s the sample size).

The flow depth within CIV was considered as the arithmetic average of the acoustic probes data for $Exp.$ 1U *Fbed*, $Exp.$ 1U *Ebed*, $Exp.$ 1D *Fbed*, $Exp.$ 1D *Ebed*, $Exp.$ 2U *Fbed* and $Exp.$ 2U *Ebed* experiments. These analyses constitute a relevant and efficient source of information to account for the uncertainty in the hydraulic conditions of each experiment. The characterization of the experimental error and propagation of uncertainty demonstrated that the variability of the hydraulic variables, particularly flow discharge, flow depth, and flow velocity, had no impact on the final results (Table 1).

The oblong bridge pier models behaved as narrow or transitional [1], depending on the resultant flow shallowness ratios (h/W). A flow shallowness of 1.13 was obtained for $Exp.$ 1D *Fbed* and $Exp.$ 1D *Ebed* (transitional piers), while values varying between 1.40 and 1.48 were attained for the remaining experiments (narrow pier category).

The flume width-to-flow depth ratios (B/h) ensured negligible effects of the secondary currents due to flume side walls on the flow. Values of 12.4, 12.3, 12.6, 12.7, 10.2 and 13.5

were obtained, respectively, for *Exp. 1U Fbed*, *Exp. 1U Ebed*, *Exp. 1D Fbed*, *Exp. 1D Ebed*, *Exp. 2U Fbed* and *Exp. 2U Ebed* (Table 1). In accordance with Yang et al. [33], that effect is solely noticeable for ratios of less than 6. The flow was, thus, free from the secondary currents at the flume mid-plane [34], where the pier models were installed. Fully developed incoming flows were obtained for all experiments, as indicated by the magnitudes of the flow and pier Reynolds number (Re and Re_D) in Table 1. The values of Froude number (Fr) lower than unity confirms the subcritical flow for experiments.

Table 1. Hydraulic variables of the fixed bed experiments.

Experiment	*Pier* (-)	Q (m^{3}s^{-1})	h (m)	V (ms^{-1})	Fr (−)	Re (−)	Re_D (−)
Exp. 1U Fbed	*Pier 11*	0.0927	$0.1610 \pm 3 \times 10^{-6}$	$0.2882 \pm 6 \times 10^{-6}$	$0.1825 \pm 8 \times 10^{-6}$	46,200 ± 0.06	31,600 ± 0.7
Exp. 1U Ebed	*Pier 11*	0.0925	$0.1628 \pm 3 \times 10^{-6}$	$0.2849 \pm 4 \times 10^{-6}$	$0.1784 \pm 4 \times 10^{-6}$	46,200 ± 0.02	31,200 ± 0.4
Exp. 1D Fbed	*Pier 14*	0.0927	$0.1588 \pm 5 \times 10^{-6}$	$0.2921 \pm 9 \times 10^{-6}$	$0.1875 \pm 1 \times 10^{-6}$	46,200 ± 0.1	32,000 ± 1.0
Exp. 1D Ebed	*Pier 14*	0.0925	$0.1579 \pm 3 \times 10^{-6}$	$0.2937 \pm 5 \times 10^{-6}$	$0.1896 \pm 6 \times 10^{-6}$	46,200 ± 0.05	32,200 ± 0.5
Exp. 2U Fbed	*Pier 11*	$0.1244 \pm 5 \times 10^{-7}$	$0.1952 \pm 1 \times 10^{-5}$	$0.3187 \pm 2 \times 10^{-5}$	$0.1665 \pm 2 \times 10^{-5}$	61,900 ± 0.2	34,900 ± 2.0
Exp. 2U Ebed	*Pier 11*	$0.1237 \pm 4 \times 10^{-7}$	$0.1486 \pm 5 \times 10^{-7}$	$0.4164 \pm 2 \text{x} 10^{-6}$	$0.2856 \pm 2 \times 10^{-6}$	61,600 ± 0.18	45,600 ± 0.2

3. Flow Field Characterization

In the present research, instantaneous velocity maps were acquired for the fixed bed experiments, before the beginning of the scouring process, at a flat bed, and at the final stage of experiments, at an eroded bed. Figure 5 provides a schematic diagram of the velocity point measurements. It comprises five half-planes, namely A, J, E, G and I. Each half-plane is composed by 4 or 5 vertical profiles, as indicated in Figure 5, where the flow velocity was measured for several points along the water depth. The position of the measuring spots varied for each application; it was dependent on flow depth and pier dimensions. In line with what was performed in the works of Kirkil et al. [35], Chang et al. [36], and Beheshti and Ataie-Ashtiani [13], among others, the measurements in each experiment were taken in the Cartesian coordinate system (x, y, z) also indicated in Figure 5. Further details can be found in Bento [28].

A minimum correlation of 70% and a minimum signal-to-noise ratio (SNR) of 15 dB were thus considered for ensuring the reliability of the instantaneous velocity data [30,31]. During the velocity measurements, the downlooking vectrino sampling frequency was chosen to be 200 Hz with a duration within the range of 120 to 300 s, depending on the magnitude of turbulence at each measuring location. These instantaneous three-dimensional velocity measurements allowed obtaining: (i) mean velocities (u, v and w); (ii) turbulent fluctuation velocities (u', v' and w'); and (iii) Reynolds shear stresses ($RSSes$) including $\overline{u'v'}$, $\overline{u'w'}$ and $\overline{v'w'}$. The derivation of such parameters provides insights of the flow magnitude and structure for the particularities of each experiment and bed morphology stage (before *vs* after scour).

Prior to data analysis, all records were visually inspected to identify potential erroneous data such as spikes, trends or abrupt discontinuities in the velocity time series. The velocity data was then processed, which streamlined the analysis of the turbulent and the time-averaged 3D velocity data, as well as the statistical and the spectral analysis of the downlooking vectrino data by computing the respective skewness and kurtosis measures for each measuring point. Results including mean values, variance, skewness and kurtosis of each of the aforementioned variables, for all experiments, can be found in Database S1 in the Supplementary Materials. For the sake of simplicity, a limited number of the most relevant plots is analysed in Section 4, namely for the different vertical profiles of the half-planes A and J. A total of 680 velocity point measurements were acquired for these two half-planes. For each vertical profile, a relative study on the characteristic flow and turbulence quantities is conducted, regarding the bridge's effective width, approach flow condition and bed configuration (flat *vs* eroded), addressed in Section 4.

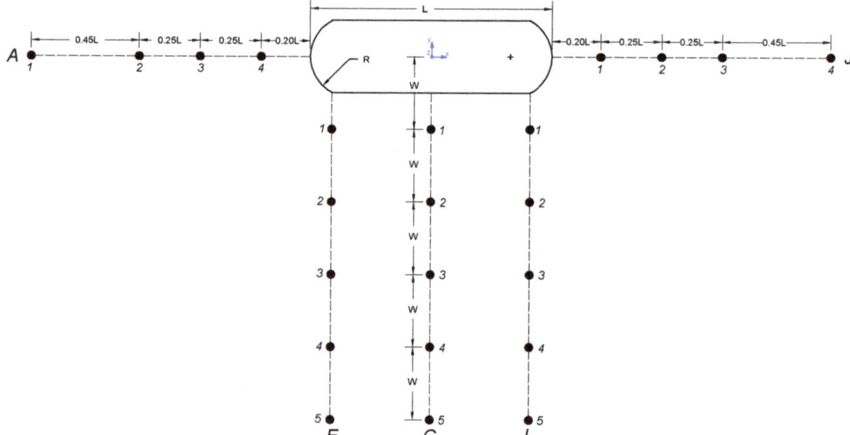

Figure 5. Schematic diagram of the velocity point measurements. Profiles on the longitudinal half-planes (*A* and *J*) are numbered from left to right, while those on the transversal half-planes (*E*, *G*, and *I*) are numbered from the pier outwards.

A characterization of the approach flow was also performed. For fully turbulent flow, over hydraulically rough sand bed surface, the dimensionless stream-wise mean velocity (u/u^*, being u^* the shear velocity), was found to follow the log-law as expected, see [37] and references therein. The roughness coefficient value adopted, k_s, was 0.00215 m (k_s = $2.5D_{50}$). The adjustment of the log-law to *Exp.* 1*U Ebed* velocity profile upstream the *Pier* 11's front, for instance, allowed obtaining the shear velocity, u^*, of 0.0174 and constant B_s = 9.74, as shown in Figure 6. These values, along with the bed Reynolds number, $u^* k_s/\nu$ $\approx$ 37, are in agreement with the literature for the hydraulically transitional boundary layer [38].

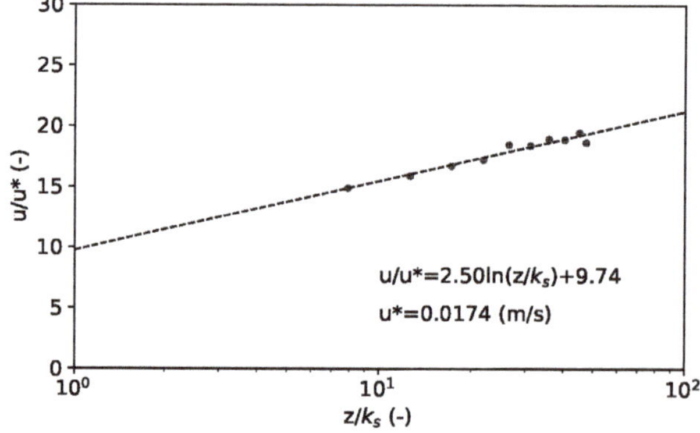

Figure 6. Longitudinal velocity profile measured in the flume mid-plane and at the distance 0.50 m upstream the *Pier* 11's front. Experimental data (○) and theoretical approximation line (dashed line).

4. Results and Discussion

4.1. Mean Velocities

Figures 7–9 depict the profiles of the velocity components at the eight mid-plane vertical profiles—*A* and *J* half-planes (Figure 5)—for the three flat bed experiments: *Exp.* 1*U Fbed*, *Exp.* 1*D Fbed* and *Exp.* 2*U Fbed*. The velocity points were scaled by a reference velocity magnitude of 0.30 ms^{-1}.

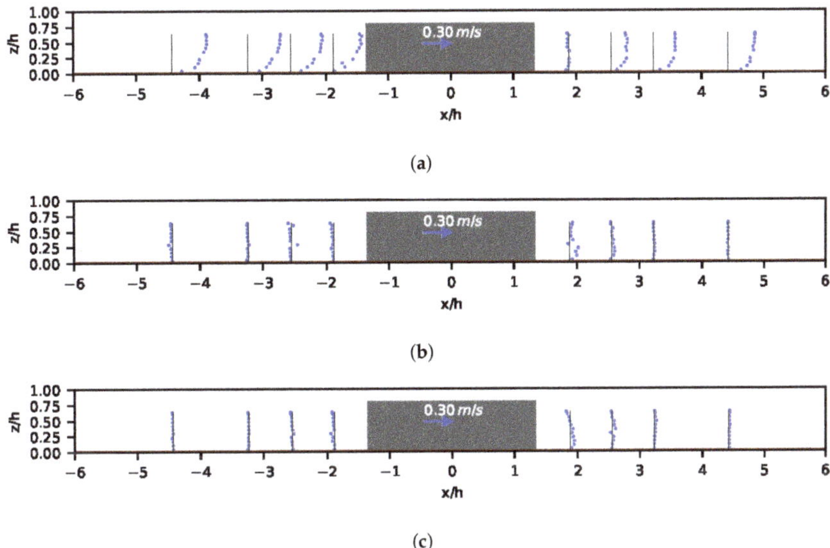

(**a**)

(**b**)

(**c**)

Figure 7. Vertical profiles of the velocity components along the pier's mid-plane for *Exp.* 1*U Fbed*. (**a**) Stream-wise component of velocity (u). (**b**) Cross-wise component of velocity (v). (**c**) Vertical component of velocity (w).

In the upstream area, the stream-wise component of the velocity (u) increases with the distance from the bed to an elevation near the flow height. Along the mid-plane surface, it decreases with proximity to the oblong bridge pier model, due to the effect of velocity reduction close to the pier's blocking effect (Figures 7a, 8a and 9a). The flat bed experiments, *Exp.* 1*U Fbed*, *Exp.* 1*D Fbed* and *Exp.* 2*U Fbed*, exhibit nearly the same pattern. The u vertical profile is diminished for *Exp.* 1*D Fbed* when compared with the *Exp.* 1*U Fbed* experiment. This can be attributed to large flow shallowness of *Exp.* 1*U Fbed* (h/W = 1.45) when compared with the correspondent value for *Exp.* 1*D Fbed* (h/W = 1.14). The higher mean approach flow velocity of *Exp.* 2*U Fbed* (in the order of 20%) justifies an increment of approximately 34% of the mean values of each of the four vertical profiles upstream the bridge pier model (*Pier* 11).

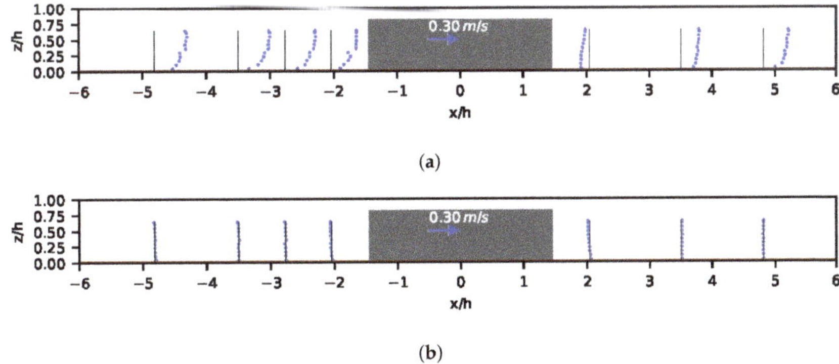

(**a**)

(**b**)

Figure 8. Vertical profiles of the velocity components along the pier's mid-plane for *Exp.* 1*D Fbed*. (**a**) Stream-wise component of velocity (u). (**b**) Vertical component of velocity (w).

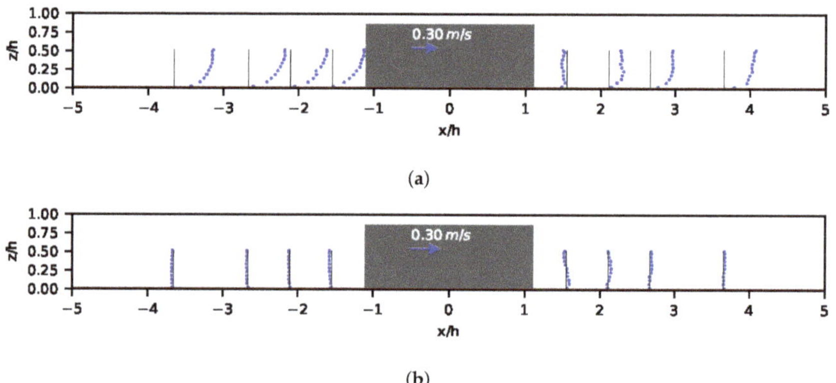

(a)

(b)

Figure 9. Vertical profiles of the velocity components along the pier's mid-plane for *Exp.* 2*U Fbed*. (**a**) Stream-wise component of velocity (u). (**b**) Vertical component of velocity (w).

In *Exp.* 1*U Fbed*, the maximum stream-wise component of velocity, u, occurred for an elevation (z) equal to 0.097 m in all the vertical profiles at the upstream side of the oblong pier (*Pier* 11); the magnitude reached a maximum value of 0.377 ms^{-1} (skewness and kurtosis coefficients of −0.70 and 1.02, respectively) for the vertical profile "1" of the A half-plane (situated 0.5 m upstream from the pier front edge, Figure 4), whereas a magnitude of 0.313 ms^{-1} (skewness and kurtosis coefficients of 2.55 and 39.43, respectively) was attained for the vertical profile "4" of the A half-plane (at roughly 0.09 m upstream from the pier front edge. For *Exp.* 2*U Fbed*, those values registered a maximum of 0.509 ms^{-1} and 0.422 ms^{-1} for the vertical profiles "1" and "4" of A, respectively. The corresponding computed skewness and kurtosis coefficients were −0.94 and 2.31 for vertical profile "1" and −1.99 and 17.42 for vertical profile "4").

Statistical moments, such as skewness and kurtosis coefficients, differing from zero means that the temporal distributions of u (for the previously mentioned example) were slightly skewed and sharpened, especially for the points measured at the vertical profile "4". For the corresponding velocity-point measurements, the v statistical moments were closer to zero (see Database S1: Flow field database). Even though observed deviations from the Gaussian distribution [39,40] are relatively small, they may reflect significant features of the large-scale turbulence dynamics. According to Nikora et al. [39] and references therein, large-scale vortices or coherent structures enhance the kurtosis coefficient, which is also corroborated by the aforementioned statistical results. Further comparisons can be undertaken for the remaining experiments.

The stream-wise component of velocity in the downstream region of the pier models presents a similar pattern between the three experiments: a general increase of u as the vertical profiles move away from the oblong bridge pier. However, it should be noted that for profile "1" of the J half-plane (see Figures 7a, 8a and 9a), in all three flat bed experiments, alternate positive and negative values were observed along the vertical profile, which can be attributed to the effect of the shedding vortexes therein developed. In line with Figure 5, the vertical profiles "1" of J half-planes are situated at 0.087 m and 0.093 m from the downstream end of *Pier* 11 and *Pier* 14, respectively.

The time average v values at the symmetry plane (y = 0) are nearly zero, as expected (Figure 7b) and reported in previous research, such as Beheshti and Ataie-Ashtiani [13] and Pandey et al. [41]. For that reason, the v vertical distributions from the remaining flat bed experiments (*Exp.* 1*D Fbed* and *Exp.* 2*U Fbed*) are not herein depicted.

The vertical profiles of the w—component velocity, represented in Figures 7c, 8b and 9b, illustrate the downward movement of flow (downflow) in front of the respective oblong bridge pier models. The w component is highest for *Exp.* 2*U Fbed*, as compared to the other two experiments (*Exp.* 1*U Fbed* and *Exp.* 1*D Fbed*, respectively). In those, the smaller values are attributed to the areas with less scouring near the pier as the stream bed surface did not permit the flow acceleration in the downward direction of the pier model. These results are in line with previous studies such as Dey and Raikar [8].

Figures 10–12 illustrate the distribution of the velocity components at the same eight mid-plane vertical profiles, from the *A* and *J* half-planes (as previously performed), for the three eroded bed experiments: *Exp.* 1*U Ebed*, *Exp.* 1*D Ebed* and *Exp.* 2*U Ebed*. Similarly, the velocity points were scaled by a reference velocity magnitude of 0.30 ms^{-1}, except for *Exp.* 2*U Ebed*, where a value of 0.40 ms^{-1} was considered.

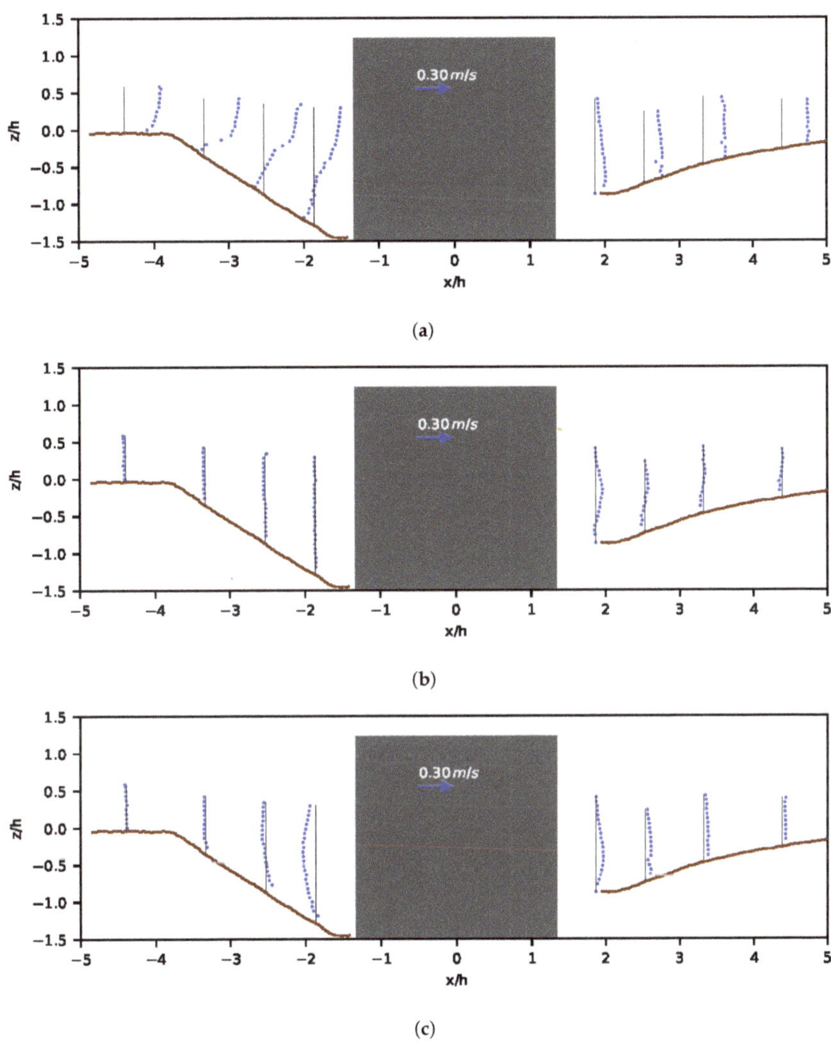

(**a**)

(**b**)

(**c**)

Figure 10. Vertical profiles of the velocity components along the pier's mid-plane for *Exp.* 1*U Ebed*. (**a**) Stream-wise component of velocity (u). (**b**) Cross-wise component of velocity (v). (**c**) Vertical component of velocity (w).

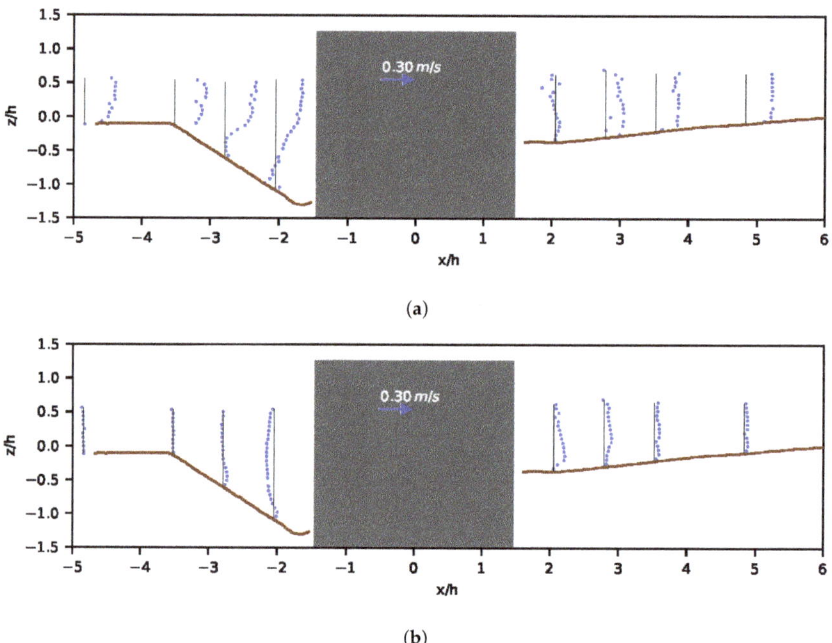

Figure 11. Vertical profiles of the velocity components along the pier's mid-plane for *Exp. 1D Ebed*. (**a**) Stream-wise component of velocity (u). (**b**) Vertical component of velocity (w).

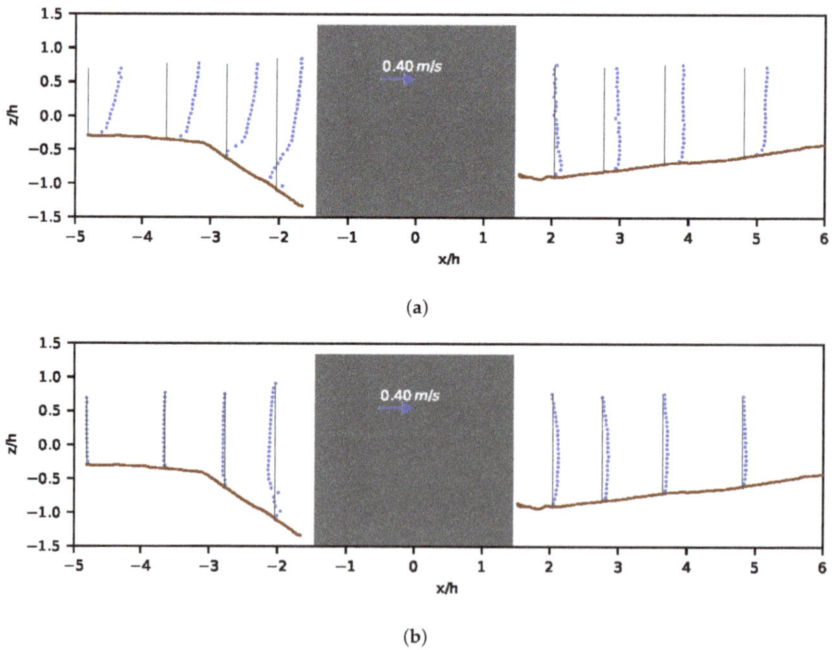

Figure 12. Vertical profiles of the velocity components along the pier's mid-plane for *Exp. 2U Ebed*. (**a**) Stream-wise component of velocity (u). (**b**) Vertical component of velocity (w).

Unlike in other flow field experimental studies, see [21] and references therein, the velocity measurements were also allowed within the scour hole of the eroded bed experiments, which explains the negative values for elevations z below the initial flat bed level (see Figures 10–12). As can be observed, vertical profile "1" of A is situated outside the upstream border of the scour hole in all experiments. The remaining vertical profiles of A and J half-planes are placed within the eroded scour holes.

The stream-wise velocity (u) profiles are marked by negative velocity within the developed scour holes, which indicates the occurrence of flow separation and recirculation, as can be seen in Figures 10a, 11a and 12a. Besides the roughly same approach hydraulic conditions of *Exp.* 1*U Ebed* and *Exp.* 1*D Ebed*, the corresponding flow shallowness values fit into distinct pier categories; the pier models behave as narrow pier for *Exp.* 1*U Ebed* and transitional pier category for *Exp.* 1*D Ebed*. As initially stated by Raudkivi [42] and corroborated by several other researchers ([43], and references therein), the parameter h/W affects the vorticity, which is associated with the generation of scour. For decreasing flow shallowness values, the influence of the surface roller on the bed increases, damping the vortices in front of the pier and consequently explaining the slightly lower u values.

A region of reversal flow was noticeable at vertical profile "4" of the A half-plane (immediately upstream from the pier), then it moved further upstream from the pier towards the scour hole edge. The u magnitudes generally increase with larger scour hole dimensions; thus, higher values were obtained for *Exp.* 2*U Ebed*. This can be explained that as the scour evolves upstream, the bed slope increases due to the formation of the scour hole, and the flow velocity increases correspondingly [21]. In Figures 10a, 11a and 12a, the downstream vertical profiles of u (from J half-plane in Figure 5) are characterized by increasing magnitudes as they move away from the oblong bridge pier models.

As performed for *Exp.* 1*U Fbed*, the v profile of *Exp.* 1*U Ebed* (shown in Figure 10b) was analysed and a similar behaviour (showing negligible changes) was observed. Hence, the remaining v profiles were not represented.

The vertical profiles of the w-component velocity illustrate the downward movement of flow (downflow) in front of the pier (attaining its maximum value for vertical profile "4" of A half-plane in all eroded bed experiments). The downward vertical velocity increased with the scour hole dimension, registering the maximum value for *Exp.* 2*U Ebed*. This finding is consistent with the results from the measurement of Dey and Raikar [8] and Li et al. [21]. Downstream of the oblong bridge pier models, w decreases with x/h for all experiments (*Exp.* 1*U Ebed*, *Exp.* 1*D Ebed* and *Exp.* 2*U Ebed*), thus reducing the lifting action. This pattern is more pronounced for the eroded bed experiments (with developed scour holes) than for the corresponding flat bed configurations.

To sum up, the mean flow velocities for the six fixed experiments indicate that with the developed scour hole, the downward flow in front of the oblong bridge pier model becomes more prominent, and the flow becomes more turbulent due to the effect of a turbulent horseshoe vortex within the scour hole, as documented by previous researchers [21,43], among others.

4.2. Reynolds Shear Stresses

The time-averaged Reynolds shear stresses, $\overline{u'v'}$, $\overline{u'w'}$ and $\overline{v'w'}$, are depicted in Figures 13–15 for experiments *Exp.* 1*U Fbed*, *Exp.* 1*D Fbed* and *Exp.* 2*U Fbed*, respectively.

The comparative study of $\overline{u'v'}$, $\overline{u'w'}$ and $\overline{v'w'}$ along the pier's mid-plane for the eroded bed experiments was also performed, as illustrated in Figures 16–18 for *Exp.* 1*U Ebed*, *Exp.* 1*D Ebed* and *Exp.* 2*U Ebed*, respectively.

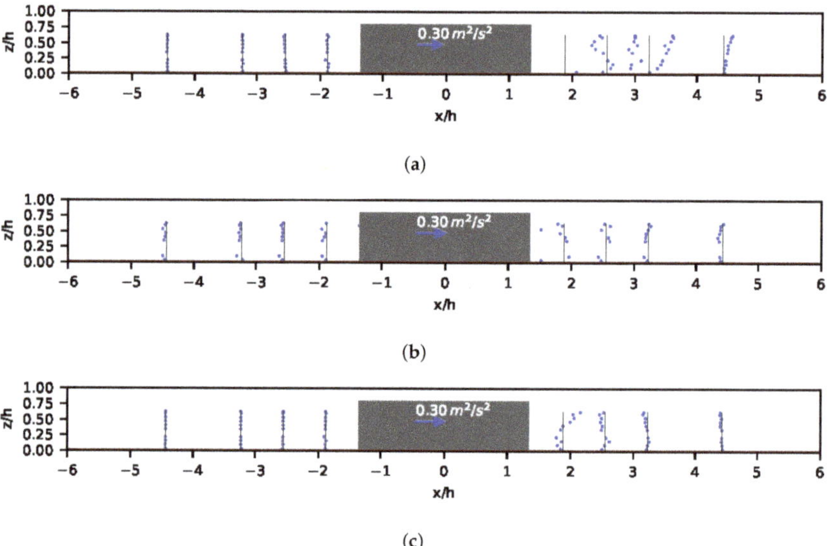

Figure 13. Vertical profiles of the Reynolds shear stresses along the pier's mid-plane for *Exp.* 1*U Fbed*. (**a**) Reynolds shear stresses ($\overline{u'v'}$). (**b**) Reynolds shear stresses ($\overline{u'w'}$). (**c**) Reynolds shear stresses ($\overline{v'w'}$).

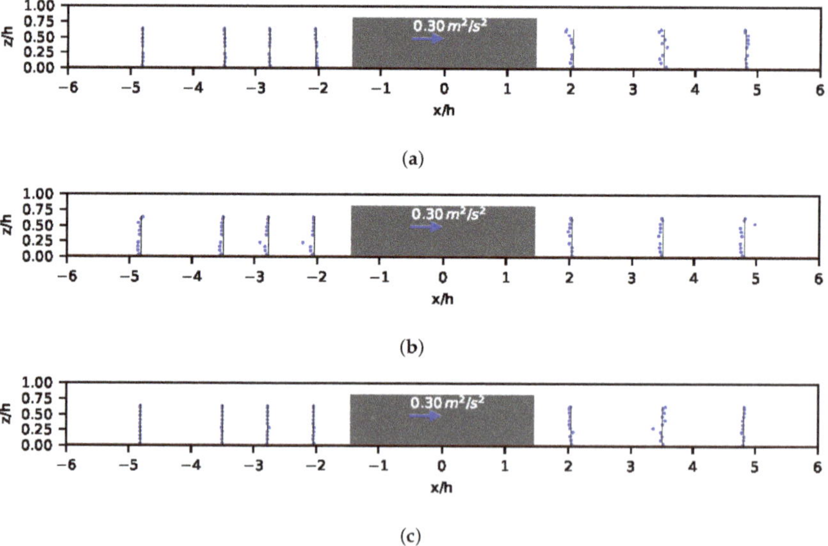

Figure 14. Vertical profiles of the Reynolds shear stresses along the pier's mid-plane for *Exp.* 1*D Fbed*. (**a**) Reynolds shear stresses ($\overline{u'v'}$). (**b**) Reynolds shear stresses ($\overline{u'w'}$). (**c**) Reynolds shear stresses ($\overline{v'w'}$).

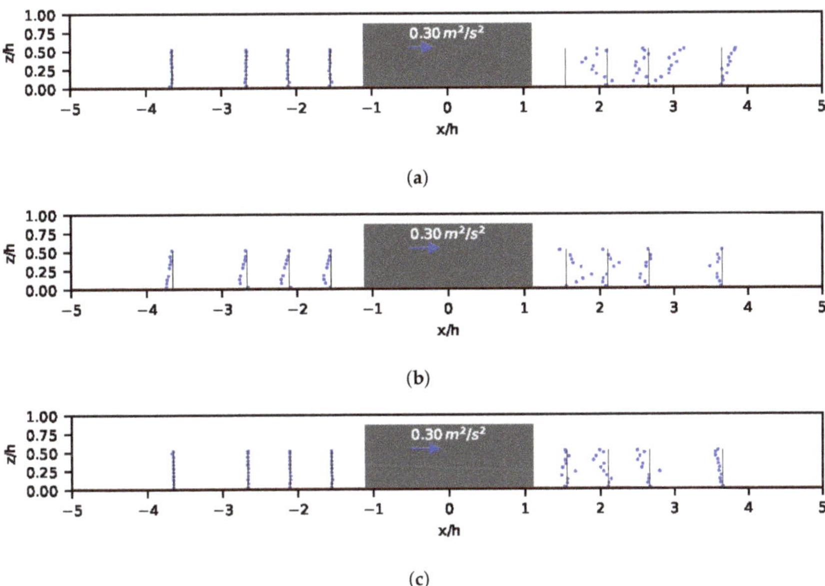

(a)

(b)

(c)

Figure 15. Vertical profiles of the Reynolds shear stresses along the pier's mid-plane for *Exp. 2U Fbed*. (**a**) Reynolds shear stresses ($\overline{u'v'}$). (**b**) Reynolds shear stresses ($\overline{u'w'}$). (**c**) Reynolds shear stresses ($\overline{v'w'}$).

Bearing in mind that the Reynolds shear stresses are linked to the capacity of the stream to transport sediments and indicate the exchange of turbulent momentum between layers of a stream [44,45], it is understandable that such capacity increases with the developing scour holes, and the region with large value enlarges and moves upstream of the correspondent scour hole.

According to Figures 13–18, no significant changes were observed in the vertical profiles of Reynolds shear stresses for different bed configurations (flat *vs* eroded). In the downstream region, the magnitudes of $\overline{u'v'}$ are greater than the corresponding points at the upstream side of the bridge pier model at both bed configurations (flat *vs* eroded). This behaviour can be attributed to the turbulent mix of fluid as a result of vortical flow in the rear side of the pier. For vertical profiles of the *J* half-plane (Figure 5), the complex flow is distinctly identified by the saw-toothed trend of the Reynolds shear stress $\overline{u'w'}$, magnified for the eroded fixed bed experiments. The values of $\overline{v'w'}$ are generally less pronounced than the other components of the Reynolds shear stress, $\overline{u'v'}$ and $\overline{u'w'}$. These results, from Figure 13–18, demonstrate that the usual assumption of isotropic turbulence is not exactly true, although it is most used as the basis of much of the theoretical analysis of turbulent flow.

Therefore, the location of the regions of the maximum Reynolds shear stresses ($\overline{u'w'}$ vertical profiles) occurs at the interfaces of the negative and positive velocity components, meaning that the velocity in this region can be very unstable. This pattern was also observed in Guan et al. [46], where the turbulent flow field, in a developing clear water (with a flow intensity of 0.53), was quantified at a circular pier.

In spite of several studies discussing the individual trends of each component of the Reynolds shear stresses, there are few authors that argue that on the mid-plane (i.e., $y = 0$), the Reynolds shear stresses involving the lateral fluctuations v' (i.e., $\overline{u'v'}$ and $\overline{v'w'}$) must be zero for symmetry reasons. In the present study, those components differed from zero mainly for the vertical profiles of the *J* half-plane. Nevertheless, this issue has not been yet thoroughly understood in the literature. It seems that theoretically the non-zero values are solely related with a signal acquiring time, of the instantaneous velocity components, not long enough to obtain reliable statistics. Further and detailed investigations regarding that issue are still needed.

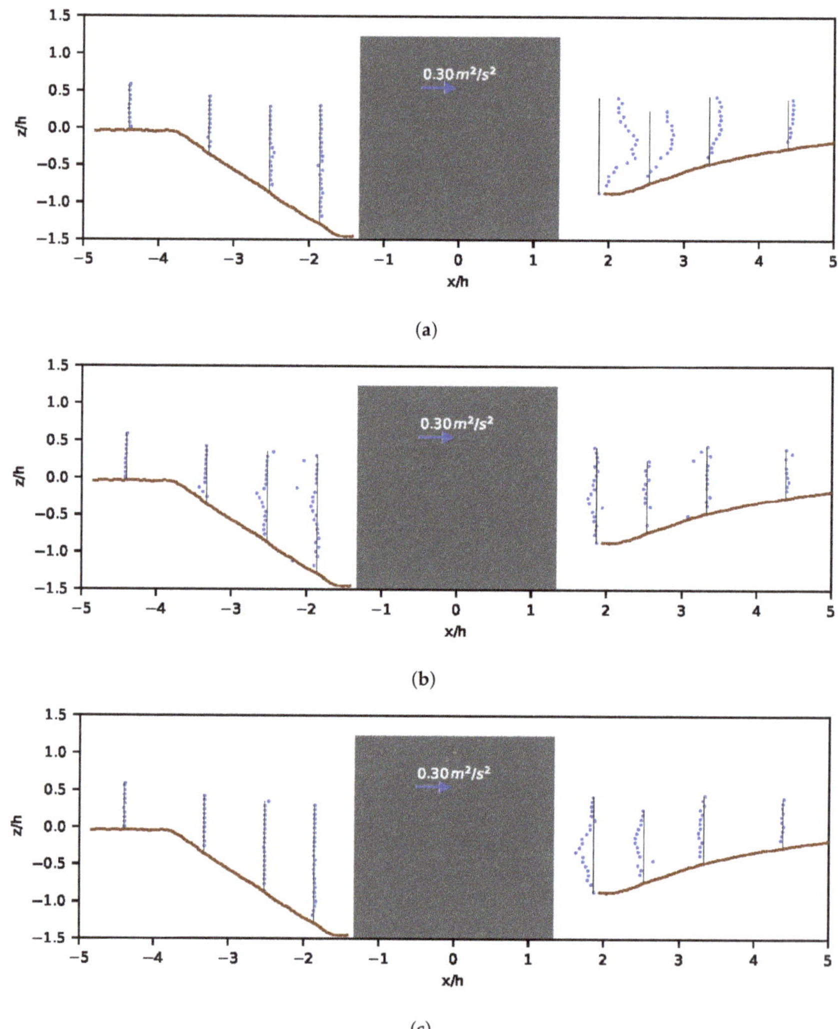

(**a**)

(**b**)

(**c**)

Figure 16. Vertical profiles of the Reynolds shear stresses along the pier's mid-plane for *Exp.* 1*U Ebed*. (**a**) Reynolds shear stresses ($\overline{u'v'}$). (**b**) Reynolds shear stresses ($\overline{u'w'}$). (**c**) Reynolds shear stresses ($\overline{v'w'}$).

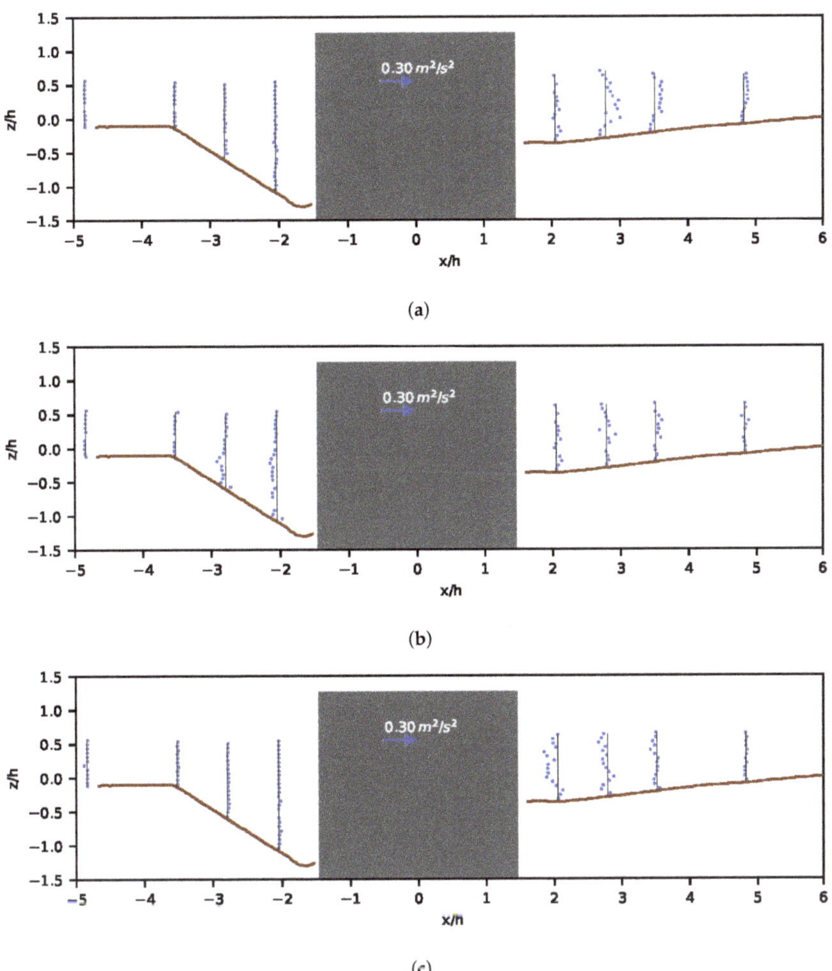

Figure 17. Vertical profiles of the Reynolds shear stresses along the pier's mid-plane for *Exp. 1D Ebed*. (**a**) Reynolds shear stresses ($\overline{u'v'}$). (**b**) Reynolds shear stresses ($\overline{u'w'}$). (**c**) Reynolds shear stresses ($\overline{v'w'}$).

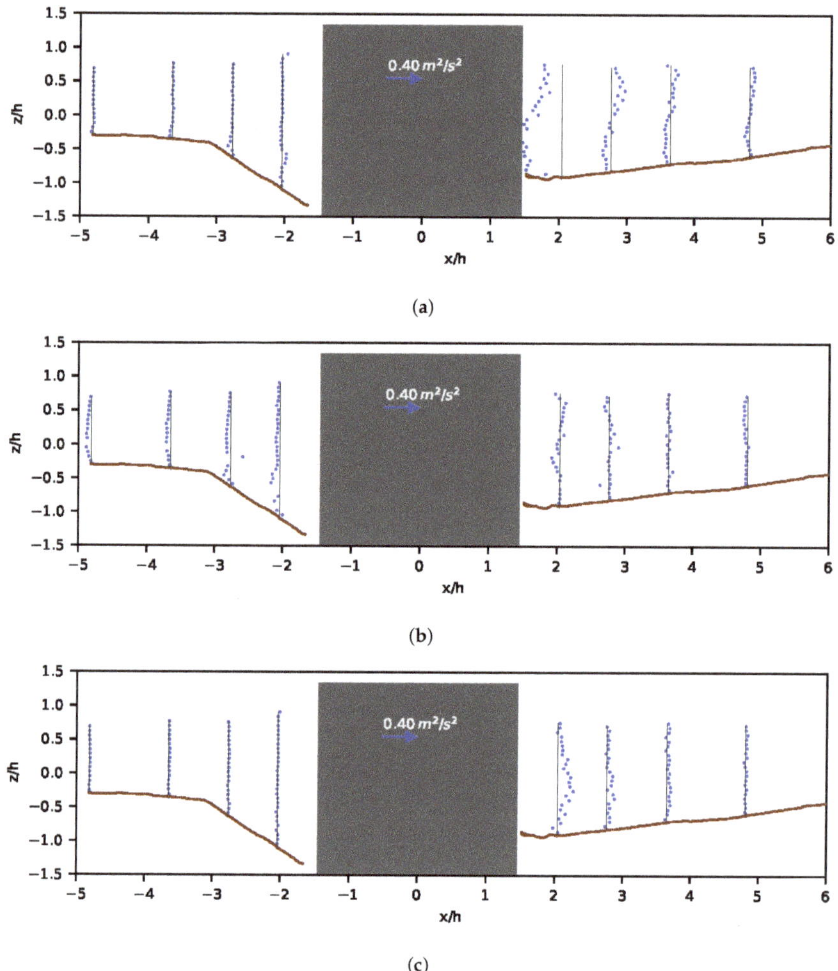

Figure 18. Vertical profiles of the Reynolds shear stresses along the pier's mid-plane for *Exp. 2U Ebed*. (**a**) Reynolds shear stresses ($\overline{u'v'}$). (**b**) Reynolds shear stresses ($\overline{u'w'}$). (**c**) Reynolds shear stresses ($\overline{v'w'}$).

5. Conclusions

This paper investigates the turbulent flow field around oblong bridge pier models in a well-controlled laboratory environment, for understanding the mechanisms of flow responsible for current-induced scour. Two different bed morphology configurations, namely (i) at the beginning of the scouring process (flat bed) and (ii) at the final stage of the scouring experiments (eroded bed) of the three local scouring experiments were herein considered. A total of six fixed bed experiments at the vicinity of the oblong bridge pier models of distinctive effective width and corresponding flow shallowness were carried out.

Measurements of the instantaneous stream-wise, cross-wise and vertical components of the flow velocity were performed by using a high-resolution acoustic velocimeter (down-looking vectrino) at eight mid-plane vertical profiles. The results showed that the presence of the developed scour hole, by creating extra flow resistance and augmenting the velocity difference within the four vertical profiles at the upstream side of the pier, changes the velocity distribution pattern and the Reynolds shear stresses. At the rear part of the pier, significant differences in flow field patterns, due to the presence of the developed scour holes, were also observed. Their magnitudes were in accordance with the approach flow conditions, such as the corresponding depth and velocity.

An important contribution was made to the understanding of the physics of the flow and turbulence characteristics with respect to the change in pier effective width, approach hydraulic conditions and bed morphology stages. The derivation of mean velocities and Reynolds shear stresses, around the oblong bridge pier models, contributes to a comprehensive description of the flow magnitude and structure for the particularities of each experiment and bed morphology stage (flat *vs* eroded). The collected data may also constitute a basis for the calibration and validation of numerical models, frequently used to characterize the scouring phenomenon, such as Computational Fluid Dynamics (CFD) tools.

In future work, a spectra analysis of the velocity measurements, and corresponding Reynolds shear stresses, at several points, both at the upstream and downstream sides of the oblong bridge pier models, can be performed. Such analysis aids the evaluation of the shedding frequency of the large scale coherent structures and further understands the turbulent dynamics, and the inertial sub-ranges of turbulence due to the scouring phenomena.

Supplementary Materials: The following are available online at www.mdpi.com/xxx/s1, Database S1: Flow field database.

Author Contributions: Conceptualization, A.M.B., T.V., J.P.P. and L.C.; methodology and design, A.M.B., T.V., J.P.P. and L.C.; formal analysis, A.M.B.; investigation, A.M.B.; writing—original draft preparation, A.M.B.; writing—review and editing, T.V., J.P.P. and L.C. All authors have read and agreed to the published version of the manuscript.

Funding: This research was funded by the Portuguese Foundation for Science and Technology (FCT) grants number PD/BD/127798/2016, UIDB/04423/2020 and UIDP/04423/2020.

Institutional Review Board Statement: Not applicable.

Informed Consent Statement: Not applicable.

Data Availability Statement: The data presented in this study are available on reasonable request from the authors.

Acknowledgments: The first author thanks the financial support of the Portuguese Foundation for Science and Technology (FCT), through the Ph.D. scholarship PD/BD/127798/2016. This research was co-supported by strategical funding from FCT UIDB/04423/2020 and UIDP/04423/2020. We would also like to thank the team of the Laboratory of Construction and Modelling of the National Laboratory of Civil Engineering (LNEC) for the technical assistance in the experimental work.

Conflicts of Interest: The authors declare no conflict of interest.

Abbreviations

The following abbreviations are used in this manuscript:

B	Flume width
B_s	Constant of log-law profile
D	Pier Model (*Pier* 14)
D_{50}	Median grain size diameter
Fr	Froude number
g	Gravity acceleration
h	Flow depth
K_s	Roughness coefficient
L	Oblong bridge pier length
n_s	Sample size
Q	Flow discharge
R	Pier semi-cylindrical surface ratio
Re	Flow Reynolds number

Re_D	Pier Reynolds number
U	Pier Model (*Pier* 11)
u	Stream-wise velocity component
u'	Stream-wise velocity fluctuation
u^*	Shear velocity
V	Approach flow velocity
v	Cross-wise velocity component
v'	Cross-wise velocity fluctuation
W	Oblong bridge pier width
w	Vertical velocity component
w'	Vertical velocity fluctuation
x	Longitudinal axis
y	Transversal axis
z	Vertical axis
ν	Fluid kinematic viscosity
ρ_s	Sediment density
σ	Standard deviation
σ_D	Standard deviation of sediment particle sizes
ADV	Acoustic Doppler Velocimeter
CFD	Computational Fluid Dynamics
$LNEC$	National Laboratory of Civil Engineering
LDV	Laser Doppler Velocimetry
PIV	Particle Image Velocimetry
$RSSes$	Reynolds Shear Stresses ($\overline{u'v'}$, $\overline{u'w'}$ and $\overline{v'w'}$)
SNR	Signal-to-Noise Ratio

References

1. Melville, B.W.; Coleman, S.E. *Bridge Scour*; Water Resources Publications, LLC: Highlands Ranch, CO, USA, 2000.
2. Arneson, L.; Zevenbergen, L.; Lagasse, P.; Clopper, P. *Evaluating Scour at Bridges*; Technical Report; Federal Highway Administration: Fort Collins, CO, USA, 2012.
3. Deng, L.; Wang, W.; Yu, Y. State-of-the-Art review on the causes and mechanisms of bridge collapse. *J. Perform. Constr. Facil.* **2016**, *30*, 04015005. [CrossRef]
4. Flint, M.M.; Fringer, O.; Billington, S.L.; Freyberg, D.; Diffenbaugh, N.S. Historical analysis of hydraulic bridge collapses in the Continental United States. *J. Infrastruct. Syst.* **2017**, *23*, 04017005. [CrossRef]
5. Proske, D. *Bridge Collapse Frequencies Versus Failure Probabilities*; Springer: Berlin, Germany, 2018.
6. Ettema, R.; Constantinescu, G.; Melville, B.W. Flow-field complexity and design estimation of pier-scour depth: Sixty years since Laursen and Toch. *J. Hydraul. Eng.* **2017**, *143*, 03117006. [CrossRef]
7. Dargahi, B. The turbulent flow field around a circular cylinder. *Exp. Fluids* **1989**, *8*, 1–12. [CrossRef]
8. Dey, S.; Raikar, R.V. Characteristics of horseshoe vortex in developing scour holes at piers. *J. Hydraul. Eng.* **2007**, *133*, 399–413.:4(399). [CrossRef]
9. Kumar, A.; Kothyari, U.C.; Raju, K.G.R. Flow structure and scour around circular compound bridge piers—A review. *J. Hydro-Environ. Res.* **2012**, *6*, 251–265. [CrossRef]
10. Melville, B.W.; Raudkivi, A.J. Flow characteristics in local scour at bridge piers. *J. Hydraul. Res.* **1977**, *15*, 373–380. [CrossRef]
11. Dey, S. Three-dimensional vortex flow field around a circular cylinder in a quasi-equilibrium scour hole. *Sadhana* **1995**, *20*, 871–885. [CrossRef]
12. Beheshti, A.A.; Ataie-Ashtiani, B. Experimental study of three-dimensional flow field around a complex bridge pier. *J. Eng. Mech.* **2010**, *136*, 143–154. [CrossRef]
13. Beheshti, A.A.; Ataie-Ashtiani, B. Scour hole influence on turbulent flow field around complex bridge piers. *Flow Turbul. Combust.* **2016**, *97*, 451–474. [CrossRef]
14. Devenport, W.J.; Simpson, R.L. Time-depeiident and time-averaged turbulence structure near the nose of a wing-body junction. *J. Fluid Mech.* **1990**, *210*, 23–55. [CrossRef]
15. Said, N.M.; Mhiri, H.; Bournot, H.; Le Palec, G. Experimental and numerical modelling of the three-dimensional incompressible flow behaviour in the near wake of circular cylinders. *J. Wind Eng. Ind. Aerodyn.* **2008**, *96*, 471–502. [CrossRef]
16. Apsilidis, N.; Diplas, P.; Dancey, C.L.; Bouratsis, P. Time-resolved flow dynamics and Reynolds number effects at a wall–cylinder junction. *J. Fluid Mech.* **2015**, *776*, 475–511. [CrossRef]
17. Jenssen, U.; Manhart, M. Flow around a scoured bridge pier: A stereoscopic PIV analysis. *Exp. Fluids* **2020**, *61*, 1–18. [CrossRef]
18. Ataie-Ashtiani, B.; Aslani-Kordkandi, A. Flow field around single and tandem piers. *Flow Turbul. Combust.* **2013**, *90*, 471–490. [CrossRef]

19. Link, O.; González, C.; Maldonado, M.; Escauriaza, C. Coherent structure dynamics and sediment particle motion around a cylindrical pier in developing scour holes. *Acta Geophys.* **2012**, *60*, 1689–1719. [CrossRef]
20. Kirkil, G.; Constantinescu, G. Effects of cylinder Reynolds number on the turbulent horseshoe vortex system and near wake of a surface-mounted circular cylinder. *Phys. Fluids* **2015**, *27*, 075102. [CrossRef]
21. Li, J.; Yang, Y.; Yang, Z. Influence of scour development on turbulent flow field in front of a bridge pier. *Water* **2020**, *12*, 2370. [CrossRef]
22. Graf, W.; Yulistiyanto, B. Experiments on flow around a cylinder; the velocity and vorticity fields. *J. Hydraul. Res.* **1998**, *36*, 637–654. [CrossRef]
23. Roulund, A.; Sumer, B.M.; Fredsøe, J.; Michelsen, J. Numerical and experimental investigation of flow and scour around a circular pile. *J. Fluid Mech.* **2005**, *534*, 351–401. [CrossRef]
24. Vijayasree, B.; Eldho, T.; Mazumder, B. Turbulence statistics of flow causing scour around circular and oblong piers. *J. Hydraul. Res.* **2020**, *58*, 673–686. [CrossRef]
25. Pasupuleti, L.N.; Timbadiya, P.V.; Patel, P.L. Flow Field Measurements Around Isolated, Staggered, and Tandem Piers on a Rigid Bed Channel. *Int. J. Civ. Eng.* **2021**, pp. 1–18.
26. Chiew, Y.M.; Melville, B.W. Local scour around bridge piers. *J. Hydraul. Res.* **1987**, *25*, 15–26. [CrossRef]
27. Bento, A.M.; Couto, L.; Viseu, T.; Pêgo, J.P. Image-Based Techniques for the Advanced Characterization of Scour around Bridge Piers in Laboratory. Available online: https://www.e3s-conferences.org/articles/e3sconf/abs/2018/15/e3sconf_riverflow2018_05066/e3sconf_riverflow2018_05066.html (accessed on 4 September 2021)
28. Bento, A.M. Risk-Based Analysis of Bridge Scour Prediction. Ph.D. Thesis, Faculdade de Engenharia da Universidade do Porto, Porto, Portugal, 2021.
29. Lee, S.O.; Hong, S.H. Turbulence characteristics before and after scour upstream of a scaled-down bridge pier model. *Water* **2019**, *11*, 1900. [CrossRef]
30. García, C.M.; Cantero, M.I.; Niño, Y.; García, M.H. Turbulence measurements with Acoustic Doppler Velocimeters. *J. Hydraul. Eng.* **2005**, *131*, 1062–1073.:12(1062). [CrossRef]
31. Montero, V.G.G.; Romagnoli, M.; García, C.M.; Cantero, M.I.; Scacchi, G. Optimization of ADV sampling strategies using DNS of turbulent flow. *J. Hydraul. Res.* **2014**, *52*, 862–869. [CrossRef]
32. Bento, A.M.; Pêgo, J.P.; Couto, L.; Viseu, T. Assessing the flow field around an oblong bridge pier. Vectrino acquisition time sensitivity analysis. *Publ. Inst. Geophys. Pol. Acad. Sci. Geophys. Data Bases Process. Instrum.* **2021**, *434*, 131–132._pas_publs-2021-038. [CrossRef]
33. Yang, S.Q.; Tan, S.K.; Lim, S.Y. Velocity distribution and dip-phenomenon in smooth uniform open channel flows. *J. Hydraul. Eng.* **2004**, *130*, 1179–1186.:12(1179). [CrossRef]
34. Cardoso, A.H.; Graf, W.H.; Gust, G. Uniform flow in a smooth open channel. *J. Hydraul. Res.* **1989**, *27*, 603–616. [CrossRef]
35. Kirkil, G.; Constantinescu, S.G.; Ettema, R. Coherent structures in the flow field around a circular cylinder with scour hole. *J. Hydraul. Eng.* **2008**, *134*, 572–587.:5(572). [CrossRef]
36. Chang, W.Y.; Constantinescu, G.; Lien, H.C.; Tsai, W.F.; Lai, J.S.; Loh, C.H. Flow structure around bridge piers of varying geometrical complexity. *J. Hydraul. Eng.* **2013**, *139*, 812–826. [CrossRef]
37. Schlichting, H.; Gersten, K. *Boundary-Layer Theory*; Springer: Berlin/Heidelberg, 2017. [CrossRef]
38. Cardoso, A.H.; Cunha, L.V.d. *Hidráulica Fluvial*; Fundação Calouste Gulbenkian: Lisbon, Portugal, 1998.
39. Nikora, V.; Nokes, R.; Veale, W.; Davidson, M.; Jirka, G. Large-scale turbulent structure of uniform shallow free-surface flows. *Environ. Fluid Mech.* **2007**, *7*, 159–172. [CrossRef]
40. Ouro, P.; Juez, C.; Franca, M. Drivers for mass and momentum exchange between the main channel and river bank lateral cavities. *Adv. Water Resour.* **2020**, *137*, 103511. [CrossRef]
41. Pandey, M.; Sharma, P.K.; Ahmad, Z.; Singh, U.K.; Karna, N. Three-dimensional velocity measurements around bridge piers in gravel bed. *Mar. Georesources Geotechnol.* **2017**, *36*, 663–676. [CrossRef]
42. Raudkivi, A.J. Functional trends of scour at bridge piers. *J. Hydraul. Eng.* **1986**, *112*, 1–13.:1(1). [CrossRef]
43. Pizarro, A.; Manfreda, S.; Tubaldi, E. The science behind scour at bridge foundations: A review. *Water* **2020**, *12*, 374. [CrossRef]
44. Bernard, P.S.; Handler, R.A. Reynolds stress and the physics of turbulent momentum transport. *J. Fluid Mech.* **1990**, *220*, 99–124. [CrossRef]
45. Carnacina, I.; Leonardi, N.; Pagliara, S. Characteristics of flow structure around cylindrical bridge piers in pressure-flow conditions. *Water* **2019**, *11*, 2240. [CrossRef]
46. Guan, D.; Chiew, Y.M.; Wei, M.; Hsieh, S.C. Characterization of horseshoe vortex in a developing scour hole at a cylindrical bridge pier. *Int. J. Sediment Res.* **2019**, *34*, 118–124. [CrossRef]

MDPI

Article

The Hydrodynamic Moment of a Floating Structure in Finite Flowing Water

Zhen Cui [1,*], Shi-Yang Pan [1] and Yue-Jun Chen [2]

1 School of Civil and Architectural Engineering, Changzhou Institute of Technology, Changzhou 213032, China; pansy@cit.edu.cn

2 MWR Key Laboratory of Yellow River Sediment, Yellow River Institute of Hydraulic Research, Yellow River Conservancy Comission, Zhengzhou 450003, China; chen-yuejun@hhu.edu.cn

* Correspondence: cuiz@cit.edu.cn; Tel.: +86-15996281130

Abstract: The implementation of floating structures has increased with the construction of new sluices for flood control, and the hydrodynamic moment of a floating structure affects the safety and operation of that structure. Based on basic hydrodynamic theory, theoretical analysis and 121 physical model tests were conducted to study the relationships between the hydrodynamic moment and the influencing factors of floating structures, namely, the shape parameter, hydraulic conditions, and draft depth. Stepwise regression fitting based on the least squares method was performed to obtain a mathematical expression of the hydrodynamic moment, and the experimental results show that hydrodynamic factors significantly influence the hydrodynamic moment of such structures. The results predicted by the mathematical expression agree with the experimental results, and thus, the proposed expression can be used to comprehensively analyze and study the safety of a floating structure under the action of flow in finite water.

Keywords: floating structure; hydrodynamic moment; finite flowing water; physical model tests; statistical diagnosis

Citation: Cui, Z.; Pan, S.-Y.; Chen, Y.-J. The Hydrodynamic Moment of a Floating Structure in Finite Flowing Water. *Fluids* **2021**, *6*, 307. https://doi.org/10.3390/fluids6090307

Academic Editors: Jaan H. Pu and Mehrdad Massoudi

Received: 24 June 2021
Accepted: 26 August 2021
Published: 31 August 2021

1. Introduction

There is an increasing trend in the applications of floating structures in inland water conservancy projects, such as inland navigation and flood control, as well as in marine engineering [1,2]. Such structures usually span a finite area (e.g., a river) with a large length-to-width ratio and exhibit characteristics similar to those of ships, offshore platforms and sluices. However, the flow patterns of such structures are variable, and their hydrodynamic mechanisms are complex [3,4]. The hydrodynamic moment on the surface of a floating structure is generated by the action of the flow and can easily cause the structure to overturn. Consequently, studies on the stability of floating structures have focused primarily on the forces, influencing factors, and responses of such structures.

To investigate floating structures in infinite water, Lee and Hong [5] studied the wave force acting on a floating structure using physical experiments and verified a test of nonlinear waves acting on a structure via the marker and cell numerical method. It was found that installing a baffle on the structure can reduce the wave force and improve stability. Venugopal et al. [6] studied floating structures at different positions under wave action through physical model tests and found that different structure types (different depth-to-width ratios) influence the wave forces in the horizontal and vertical directions. Roy and Ghosh [7] measured the horizontal force and momentum of plates at different depths under wave action through model tests and found that the force and momentum converged with increasing wave periods and varied with the positions of the plates. Hayatdavoodi et al. [8] and Seiffert et al. [9] experimentally studied the force of solitary waves on coastal pontoons and discovered that the water depth, flow velocity, waves, and immersion depth of the pontoon affected the horizontal forces on the vessel. In addition, the vertical force decreased with increasing water level, and the experiments revealed that the interaction between waves and pontoon bridges could effectively change the force.

In contrast, the forces on and influencing factors of a floating structure in finite water are considerably different from those in infinite water. For instance, Xing et al. [10,11] found that the response of a floating structure in finite water is different from that in infinite water; specifically, the boundaries of a finite domain act to reflect water waves, and a resonance phenomenon occurs between the operating structure and the water, with the additional mass-produced being related to the width of the water domain and time. Wang et al. [12,13] applied a numerical method to study the moments of floating structures in finite flowing water and found that the flow field and obtained moment varied differently with changes in the shape, position and water level conditions of the structure. Fu and Yin et al. [14] obtained the influencing factors of the speed of a floating sluice and the relationship between the overturning characteristics and ballast using hydraulic model tests. The results showed that the flow velocity of a sluice is related to the inflow (discharge) of water and that the safety of a floating sluice can be improved by increasing the ballast or reducing the water inflow. Lu and Wang [15] studied the hydraulic characteristics and corresponding velocity characteristics of a fixed floating structure under the action of solitary waves and reported that increasing the incident wave amplitude increases the vertical forces on the structure and that the horizontal force on the structure increases with increasing draft depth. Rodrigues and Guedes Soares [16] simulated the stability of floating structures with different shapes in still water and showed that the shape of the structure has a large influence on its stability. Cui et al. [17,18] analyzed the influencing factors of the hydraulic characteristics around floating structures in finite flowing water and described the velocity distribution and the hydraulic characteristics, which were affected by the shape, position and hydraulic conditions of the structure. According to the principles of a system in equilibrium, Fu et al. [19] analyzed the hydraulic characteristics of a new type of floating sluice in a current based on hydraulic model tests and found that the stability of the structure was affected not only by its center-of-gravity position, free surface, ballast form and operating speed but also by the flow and constraint conditions.

Furthermore, Venugopal [20] experimentally studied the responses of regular and irregular floating structures under the actions of waves and currents and demonstrated that, unlike in static water, currents and the structure type affect the drag force and the inertia coefficient. Rey et al. [21] discussed floating structures under the simultaneous action of flow and waves and found that the water pressure and near-bottom velocity affect the bottoms of structures under deep water conditions and that the effects of waves and currents on the bottom of a submerged structure are not negligible. Johnson et al. [22] studied the stability of a floating breakwater under the action of waves and flow via physical tests and described the effects of the wave height and water flow velocity on the breakwater. On the basis of thin plate theory and linear potential flow, Wan et al. [23] verified the structural deformation and bending moment distribution of a floating structure via the modal function expansion method and the characteristic function expansion method. Zhang [24] systematically described the hydraulic characteristics of a floating gate operating under the action of water through experiments and numerical methods. Positive correlations were found between the rotational resistance and the draft depth, rotational speed, and flow velocity, and the pressure distribution on the water-facing surface of the gate was found to be greater than that on the downstream backwater surface. Rey and Touboul [25] studied the hydraulic load of a submerged plate under the action of water flow and discovered that the water flow influences both the reflection coefficient and the horizontal force acting on the plate but has little effect on the vertical force. Pu et al. [26,27] proposed a method to improve the measured velocity and turbulence structure data in the near-bed region through experimental and computational approaches. Li and Lin [28] developed a two-dimensional numerical calculation method under wave and current conditions and discussed the hydrodynamic coefficients corresponding to a cylindrical floating structure affected by waves and currents. The drag force was found to decrease as the wave height increased, while the inertial force decreased only slightly.

The purpose of this paper is to analyze the hydrodynamic moment characteristics of a floating structure in finite flowing water. The hydrodynamic pressure on a floating structure influences whether the structure overturns; this phenomenon motivates the development of a method to accurately calculate the overturning moment via the hydrodynamic pressure. Based on dimensional analysis, the influencing factors of the floating hydrodynamic moment are studied by conducting a large number of physical experiments with respect to the shape, flow parameters, and draft depth of the floating structure, and the variations in and characteristics of the hydrodynamic moment are obtained. In light of the experimental results, a formula of the overturning moment in terms of the hydrodynamic pressure in flowing water is given. Moreover, the statistical theory is applied to diagnose and verify the correctness of the formula to provide a basis for the safe operational control of a floating structure under the action of flow, which has important theoretical significance and practical application value.

2. Experimental Design and Configuration

When a floating structure is employed in a water conservancy project, it is used to control the water level and regulate the flow. A floating structure operates either in a submerged state (completely immersed in the water) or a floating state (partially submerged in the water). The influencing factors that affect the hydrodynamic moment and measurement methods of floating structures in these submerged and floating states are not the same. Hence, the hydrodynamic moment of a floating structure in a submerged state has previously been qualitatively and quantitatively analyzed by arranging miniature dynamic pressure sensors on the surface of the structure through the microintegral method [29]. However, because part of a floating structure in a floating state remains above the water surface, which is inherently different from the submerged state, the hydrodynamic moments of floating structures in a floating state should also be studied.

2.1. Experimental Configuration and Dynamic Moment Measurement

During the operation of a floating structure in flowing water, the factors affecting the hydrodynamic moment mainly include the shape of the floating structure, the draft depth of the structure, and the hydraulic conditions. According to the factors affecting the hydrodynamic moment in hydraulic engineering, indoor experiments were established, and the main parameter conditions were analyzed. Flume experiments involving a self-circulating water supply system were performed to study the hydraulic characteristics and patterns of the hydrodynamic moment of a floating structure; the flume was made of plexiglass with a smooth surface. The moment and hydraulic parameters were measured under steady flow conditions. The experimental configuration is presented in Figure 1a; as shown in Figure 1b, the floating structure was set in the central area of the flume. With the width of the floating structure perpendicular to the flume being 0.30 m, water flowed downstream beneath the structure rather than around it. An independently developed stepper motor and computerized numerical control (CNC) speed control system was used to control the position of the floating structure at an accuracy of 0.10 mm. The upstream and downstream water levels of the model were measured using needle water level gauges with a measurement accuracy of ±0.10 mm. Typical velocities were obtained by an acoustic Doppler velocimeter at an accuracy of 0.01 cm/s, and the acquisition frequency was 200 Hz [30,31]. The flow characteristics around the floating structure were observed along the longitudinal section.

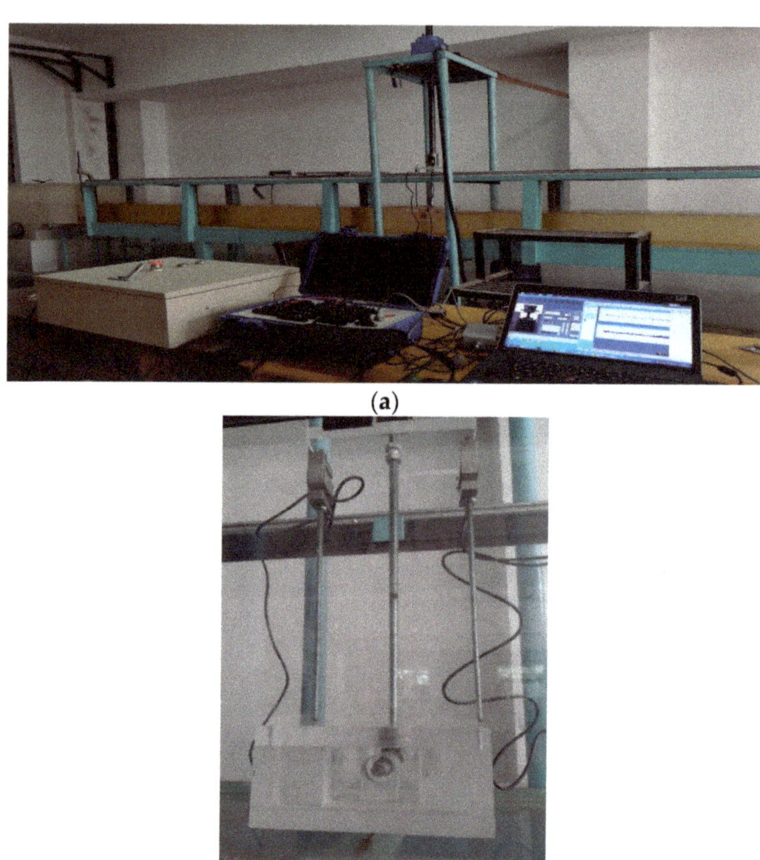

Figure 1. Diagram of the test model: (a) Panoramic view of the experimental configuration, (b) The diagram of the floating structure.

To measure the hydrodynamic moment of the structure, the fixed floating structure capable of both sinking and floating was connected to a rotating rod set at the geometric center of the structure to ensure that the structure could freely (but only slightly) rotate along the direction of the water flow, and a tension pressure sensor was arranged along the center line of water flow at each end (upstream and downstream) of the upper part of the structure. The sensor was a CY201 digital pressure and temperature transmitter with an accuracy of ±0.50‰, and the acquisition frequency was 512 Hz. When a floating structure is under the action of flow, it rotates along the direction of the flow; therefore, the hydrodynamic moment can be calculated by the pressure values output from the tension pressure sensors (see Figure 2). The pressures on the two sensors are denoted F_1 and F_2, and the distances from the center of the structure to the sensors are denoted D_1 and D_2, respectively. The hydrodynamic moment was calculated with clockwise being negative and counterclockwise being positive as follows:

$$M = F_1 D_1 - F_2 D_2 \quad (1)$$

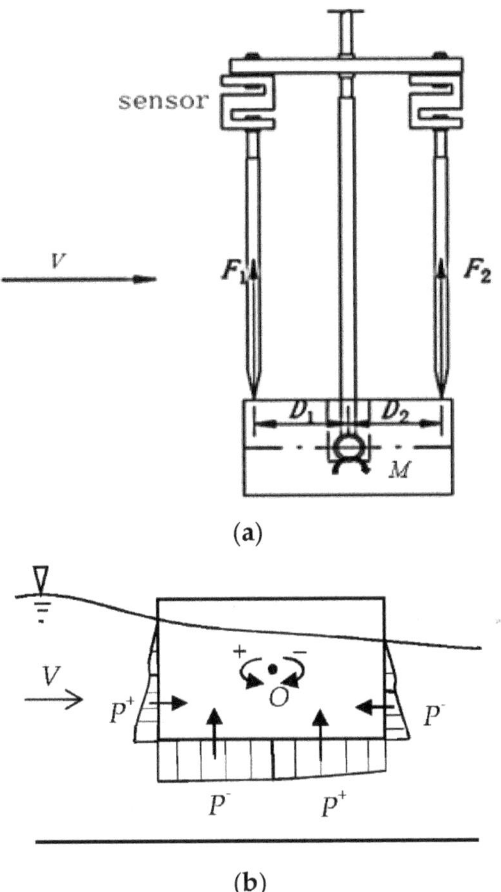

Figure 2. Schematic of moment measurement: (**a**) Layout of the sensor, (**b**) Pressure profile of the floating structure.

2.2. Dimensional Analysis and Parameter Design of the Floating Structure

The flow pattern around a floating structure is complicated, and the surrounding velocity distribution is uneven, which affects the dynamic moment and stability when the structure moves upward and downward during operation.

We define the hydrodynamic moment of a floating structure per unit length (L = 1.00 m) as $M_{Floating}$ or M_F for short. The influencing factors of M_F [MLT^{-2}] mainly include the width of the structure B ([L]), the height of the structure a ([L]), a draft of the structure h ([L]), downstream water depth H' ([L]), gravitational constant of acceleration g ([LT^{-2}]), water density ρ ([ML^{-3}]), upstream and downstream water level difference ΔH ([L]), and average velocity of the upstream section v ([LT^{-1}]). Figure 3 shows a schematic diagram of the main parameters affecting the floating structure. According to dimensional analysis, the expression of the parameters influencing the hydrodynamic moment is as follows:

$$f(M_F, B, g, \rho, a, h, \Delta H, H', v) = 0 \quad (2)$$

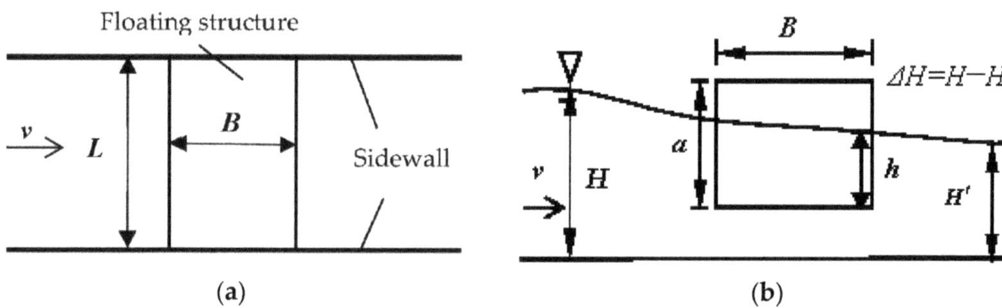

Figure 3. Main parameters of the floating structure: (**a**) Plan view, (**b**) Front view.

The downstream water depth H' ([L]) in the geometric dimension, gravitational constant of acceleration g [L/T^2] in the kinematic dimension, and water density ρ in the dynamic dimension ([ML^{-3}]) are selected as the three basic physical quantities. According to the π theorem:

$$\frac{M_F}{\rho g H'^3} = f(\frac{B}{H'}, \frac{a}{H'}, \frac{h}{H'}, \frac{\Delta H}{H'}, \text{Fr}) \tag{3}$$

where the dimensionless factor on the left side is defined as M_f. We define $\pi_2/\pi_3 = B/a$ as the aspect ratio of the floating structure; h/H' is the relative draft; $\Delta H/H'$ is the relative water level difference, and $\pi_6 = \text{Fr} = v/\sqrt{gH}$ is the Froude number of the upstream section. The length of the section is five times the water depth of the flume at the upstream end of the floating structure, where the flow is stable and is not affected by the structure. The simplified result is:

$$M_f = f(\frac{B}{a}, \frac{h}{H'}, \frac{\Delta H}{H'}, \text{Fr}) \tag{4}$$

The dimensionless hydrodynamic moment of the floating structure is related to the shape B/a, the relative draft h/H', the relative water level difference $\Delta H/H'$ of the floating structure, and the Froude number Fr.

The influencing factors of the hydrodynamic moment are the shape of the floating structure, upstream and downstream water levels, draft depth, and inflow conditions. The height of the floating structure a selected in the experiment was 10.00 cm. Four different widths of floating structure B were selected (10.00 cm, 20.00 cm, 30.00 cm, and 40.00 cm); therefore, the width-to-height aspect ratios were 1.00, 2.00, 3.00, and 4.00, respectively. The flow rates per unit width were set to 0.05 m^2/s, 0.06 m^2/s, 0.07 m^2/s, and 0.08 m^2/s; the draft depths of the structure were 2.00 cm, 4.00 cm, 6.00 cm, and 8.00 cm; and the downstream water levels were 18.00 cm, 20.00 cm, 22.00 cm, and 24.00 cm, respectively (see in Table 1). To ensure the completeness of the test, orthogonal tests were also conducted, resulting in a total of 121 test groups.

Table 1. Experimental parameters.

a/m	B/m	h/m	q/(m^2 s^{-1})	H'/m			
0.10	0.10 0.20 0.30 0.40	0.02	0.05	0.18	0.20	0.22	0.24
			0.06	0.18	0.20	0.22	0.24
			0.07	0.18	0.20	0.22	0.24
			0.08	0.18	0.20	0.22	0.24
		0.04	0.05	0.18	0.20	0.22	0.24
			0.06	0.18	0.20	0.22	0.24
			0.07	0.18	0.20	0.22	0.24
			0.08	0.18	0.20	0.22	0.24
		0.06	0.05	0.18	0.20	0.22	0.24
			0.06	0.18	0.20	0.22	0.24
			0.07	0.18	0.20	0.22	0.24
			0.08	0.18	0.20	0.22	0.24
		0.08	0.05	0.18	0.20	0.22	0.24
			0.06	0.18	0.20	0.22	0.24
			0.07	0.18	0.20	0.22	0.24
			0.08	0.18	0.20	0.22	0.24

3. Results and Discussion

Figure 4 shows a schematic diagram of the flow along the longitudinal section. Due to the existence of the floating structure, the flow from the upstream region passes under the structure, which reduces the cross-section of the water flow and increases the velocity; on the front surface of the structure, the velocity is parallel to the flow direction is zero. The water surface profile drops slightly in the upstream region of the structure, and the water level on the front surface of the floating body is higher than that on the back surface. The presence of the structure creates a recirculation zone in the downstream region, which causes the pressure on the back surface of the structure to decrease.

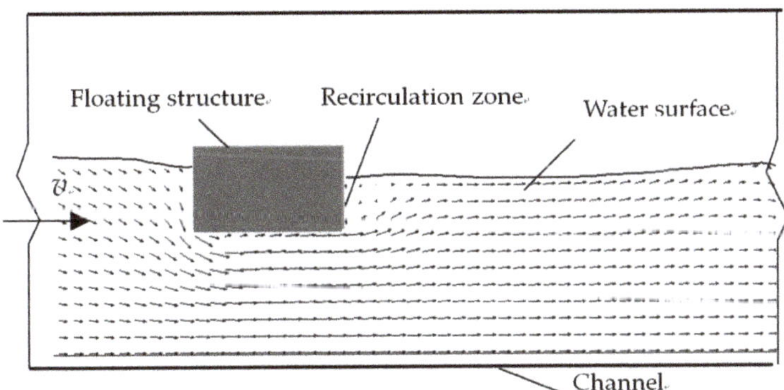

Figure 4. Schematic diagram of the flow.

3.1. Shape of the Floating Structure

Figure 5 shows a scatter diagram relating the dimensionless hydrodynamic moment M_f to the structure shape parameter B/a under the same hydraulic conditions. With increasing B/a, M_f increases. During the experiment, with increasing B/a, the flow velocity decreases significantly near the center of the structure's lower surface. The small recirculation zone near the lower center of the structure expands to the back half of the structure. When the shape parameter of the floating structure increases, the surface pressure difference caused by the uneven distribution of the dynamic pressure increases, and the distance between the action point and the center of the structure also increases, resulting in a rise in the hydrodynamic moment. For a larger structure, the pressure drop from the entrance of the bottom gap to the downstream part increases, causing an increased pressure difference and giving rise to the hydrodynamic moment.

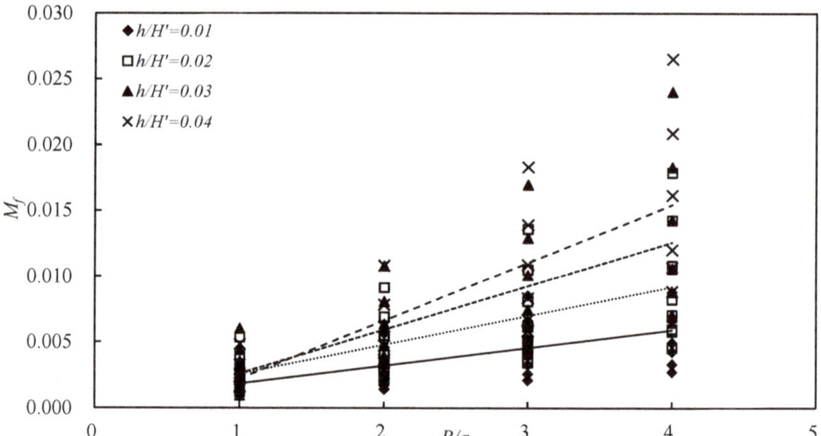

Figure 5. Relationship between M_f and the floating structure's shape.

The slope in the graph increases with increasing h/H' and shows a linear relationship. The relationship between the shape of the structure and M_f was obtained through a fitting analysis:

$$M_f = k_1 Bh/aH' + c_1 \tag{5}$$

where k_1 and c_1 are the coefficient and constant terms, respectively, of the independent variable Bh/aH'.

3.2. Draft Depth

When the downstream water depth changes, the relationship between the relative draft depth and the hydrodynamic moment is complicated. However, when the factors h/H' and $\Delta H/H'$ are combined, an obvious pattern emerges. Figure 6 shows the relationship between M_f and $h/\Delta H$, indicating that M_f has an obvious logarithmic relationship with $h/\Delta H$, and its value increases with increasing B/a. A non-uniform distribution of the flow velocity causes a change in the hydrodynamic moment at different draft depths. The hydrodynamic pressure acting on the surface of the structure decreases obviously if the draft depth is small. When the water level difference increases, the velocity around the floating structure rises, and the pressure difference between the front and back surfaces of the structure increases, resulting in an increase in the hydrodynamic moment.

In addition, there is a negative linear correlation between B/a and the coefficient term k_2. Hence, the correlation between M_f and the draft depth can be expressed as follows:

$$M_f = k_2 \frac{B}{a} \ln\left(\frac{h}{\Delta H}\right) + c_2 \tag{6}$$

where k_2 and c_2 are the coefficient and constant terms, respectively, of the independent variable $\ln(\frac{h}{\Delta H})$.

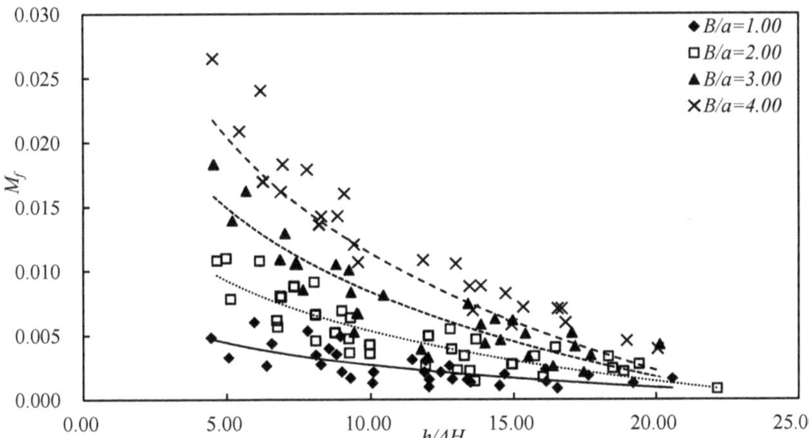

Figure 6. Relationship between M_f and the draft depth.

3.3. *Relative Water Level Difference*

Figure 7 shows the relationship between M_f and the relative water level difference $\Delta H/H'$. M_f increases with increasing $\Delta H/H'$, and when the structure is large, the growth rate of the hydrodynamic moment shows an increasing trend. This is similar to the conclusion that the moment increases with increasing *B*/*a*. When $\Delta H/H'$ increases, the pressure difference between the front and back surfaces of the structure increases. With an increasing water level difference, the hydrodynamic pressure acting on the upstream face of the floating structure is obviously higher than that acting on the downstream face; therefore, the structure tilts downstream, and the moment increases obviously. The hydrodynamic pressure acting on the downstream face was found to be relatively low due to the recirculation zone existing near the downstream face of the structure. As a consequence, the structure becomes increasingly unstable as a result of the increasing water level difference and thus can be easily unbalanced.

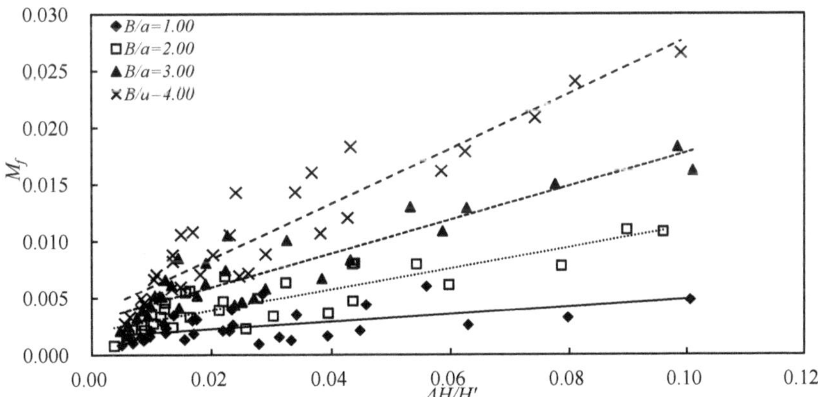

Figure 7. Relationship between M_f and $\Delta H/H'$.

There is a clear linear relationship between the different slopes and *B*/*a*. The relationship between M_f and $\Delta H/H'$ was obtained through a fitting analysis:

$$M_f = k_3 B/a \cdot \Delta H/H' + c_3 \tag{7}$$

where k_3 and c_3 are the coefficient and constant terms, respectively, of the independent variable $B\Delta H/aH'$.

3.4. Froude Number

Figure 8 shows the relationship between the hydrodynamic moment M_f and Fr^2. The M_f value of the floating structure increases with increasing Fr^2. Moreover, with increasing B/a, the growth rate of M_f tends to increase. With increasing velocity, the recirculation zone in the downstream area increases, thereby increasing the velocity difference between the upstream and downstream faces of the structure. The kinetic energy of the flow is converted into potential energy due to the water acting on the upstream face, and the dynamic pressure increases, thereby increasing the pressure difference between the front and back surfaces of the structure, which increases the hydrodynamic moment and decreases its stability.

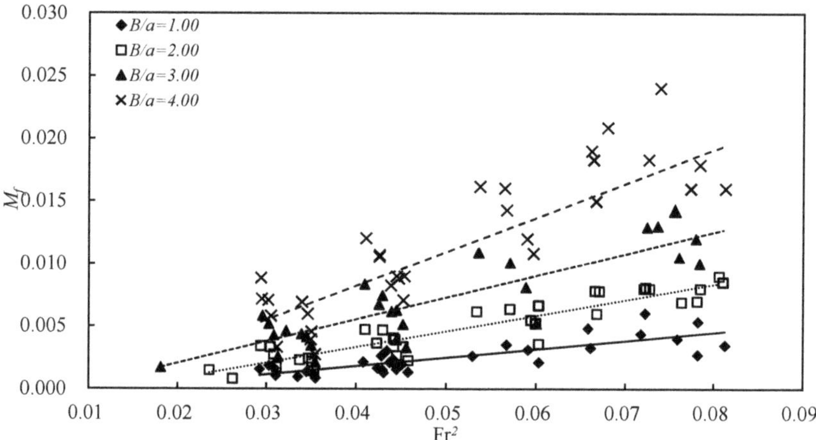

Figure 8. Relationship between M_f and Fr^2.

If we consider the effect of the Froude number and shape parameter variables as a new independent variable BFr^2/a on M_f, the following formula can be obtained:

$$M_f = k_4 BFr^2/a + c_4 \tag{8}$$

where k_4 and c_4 are the coefficient and constant terms, respectively, of the independent variable BFr^2/a.

An analysis of the influencing factors in combination with the hydrodynamic moment of the floating structure in a submerged state reveals both similarities and differences in the hydrodynamic moments of the floating structure in the two different states. In both states, as the shape parameter of the structure and the water level difference increase, the hydrodynamic moment of the floating structure shows an increasing trend. However, the variation trends of Fr and the position (draft depth) in these two states are not the same: in a submerged state, the surface of the structure is affected by the pressure of the flowing water, while in a floating state, the upper surface of the floating body is at atmospheric pressure. Hence, the area over which kinetic energy acts on the surface of the floating structure differs between the two states, resulting in a change in the pressure difference.

4. Stepwise Regression Analyses and Validation of M_f

4.1. Regression Analysis of M_f

Fitting the relevant factors of the scatter plots indicates that four dimensionless parameter-independent variables affect the hydrodynamic moment: Bh/aH', $B\mathrm{Fr}^2/a$, $B\Delta H/aH'$ and $B\ln(h/\Delta H)/a$. Based on the least squares method, stepwise regression analysis was performed to optimize and eliminate the relevant independent variables. The confidence intervals of the introduced and excluded variables are 95% and 90%, respectively. The optimal mathematical expression of M_f was obtained by excluding the repetitive variable $B\Delta H/aH'$. The results of significance analysis of the regression equation coefficients are shown in Table 2.

Table 2. Significance analysis of the coefficients.

Independent Variable	Nonstandardized Coefficients		*T*	*Sig.*	Coefficient 95% Confidence Interval		Collinear Statistics	
	Coefficient	Standard Error			Lower Limit	Upper Limit	Tolerance	*VIF*
Constant	−0.001	0.000	−1.73	0.08	−0.001	0.000		
$B\Delta H/aH'$	0.045	0.002	19.99	0.00	0.040	0.049	0.66	1.51
$B\mathrm{Fr}^2/a$	0.036	0.002	15.73	0.00	0.031	0.040	0.51	1.97
Δh	−0.001	0.000	−2.21	0.00	0.000	0.000	0.73	1.38

According to the significance of the coefficients and a collinear analysis, the formula for calculating the dimensionless M_f of a floating structure of unit length is as follows:

$$M_f = -0.001 + \frac{B}{a}\left[0.045\frac{\Delta H}{H'} + 0.036\mathrm{Fr}^2 - 0.001\ln\left(\frac{h}{\Delta H}\right)\right] \tag{9}$$

where B is the width of the floating structure [m]; h is the draft depth [m]; a is the height of the floating structure [m]; H' is the depth of the downstream water [m]; ΔH is the upstream and downstream water level difference [m], and Fr is the Froude number. The moment center of the formula is the geometric center of the floating structure. This formula is applicable under conditions of $1 \leq B/a \leq 4$, $a > h > 0$, $0 < \Delta H/H' \leq 0.10$, and $\mathrm{Fr} \leq 0.35$.

The above formula reflects the hydrodynamic moment of a floating structure from entering to becoming fully immersed in the water. When the floating structure is used as a sluice in a water conservancy project, it sinks to the sluice floor to connect with the river channel. The factors involved in this formula are the shape parameter and position of the structure and the hydraulic conditions around the floating structure. Therefore, the topographic features of the upstream and downstream parts of the channel have little influence on the hydrodynamic moment, and thus, the above formula can be applied to complex irregular channels and complex bedforms.

4.2. Verification and Error Analysis

To evaluate the accuracy of Equation (9), the adjusted multiple correlation coefficient (AMCC) and standard error of estimation (SEE) were used. The AMCC is a modified version of the multiple correlation coefficient; it gives the percentage of variation explained by only those significant variables that affect the predicted value in reality. The AMCC value indicates the goodness of the fit of M_f; thus, the AMCC was used to verify the correctness of the prediction and to ascertain whether the regression model is satisfactory. The SEE value evaluates the reliability of the data, with smaller values indicating higher reliability, which indicates that the observations are closer to the fitted line. The AMCC and SEE are defined as follows:

$$R^2 = \frac{\sum_{k=1}^{K}(\tilde{y}_k - \overline{y})^2}{\sum_{k=1}^{K}(y_k - \overline{y})^2} \tag{10}$$

$$\text{AMCC} = R^2 - \frac{J \times (1 - R^2)}{K - J - 1} \quad (11)$$

$$\text{SEE} = \sqrt{\frac{\sum_{k=1}^{K} (y_k - \widetilde{y_k})^2}{K - J - 1}} \quad (12)$$

where R^2 is the coefficient of determination, K is the size of the dataset, J is the number of dimensionless independent variables, and y_k is the measured value of M_f. Here, $\widetilde{y_k}$ is the value of M_f estimated by Equation (9), and $\overline{y}$ is the mean value of y_k.

The AMCC and SEE values determined for Equation (9) are 0.94 and 0.01, respectively, thus proving that the significance is strong.

The model was further tested to judge the applicability of the regression equation and the fitting effect. A 95% confidence interval was selected; the corresponding value of F is 454.51. In addition, the autocorrelation of the model, which reflects the ability to conduct valid statistical tests, was evaluated by using the commonly used Durbin–Watson (DW) test. The value of DW was calculated as 2.39, indicating that no autocorrelation exists in the model. The correlation coefficient of the formula between the variables is low, suggesting that the independent variables are not correlated. To eliminate the instability of the model, the variance inflation factor (VIF) was used to diagnose the existence of multicollinearity. The maximum VIF value was 1.97; thus, there is no multicollinearity problem between the variables and M_f. The F-test, DW test and VIF test formulas are given as follows:

$$F = \frac{\sum_{k=1}^{K} (\widetilde{y_k} - \overline{y_k})^2 / K}{\sum_{k=1}^{K} (y_k - \widetilde{y_k})^2 / (K - J - 1)} \quad (13)$$

$$DW = \frac{\sum_{t=1}^{K} (e_t - e_{t-1})}{\sum_{t=2}^{K} {e_t}^2} \quad (14)$$

$$VIF_i = \left(1 - R_i^2\right)^{-1} \quad (15)$$

where $\overline{y_k}$ is the mean value of $\widetilde{y_k}$, e_t is the error term at time t, and ${R_i}^2$ is the multiple coefficient of determination of the independent variable.

The standard residuals need to satisfy the requirements of randomness and normality to confirm the correctness of the obtained formula. The following results were obtained by analyzing the standard residuals. Figure 9 illustrates the corresponding statistical analysis of the residuals of Equation (9) and a histogram of the residuals with a normal probability curve; the residuals present a normal distribution. The 95% distribution of standardized residuals is between −2 and +2, which suggests that the model assumptions are reasonable.

For M_f, a comparison between the measured and calculated results is shown in Figure 10, suggesting good agreement. Thus, the development of a calculation formula for the overturning moment was deemed successful.

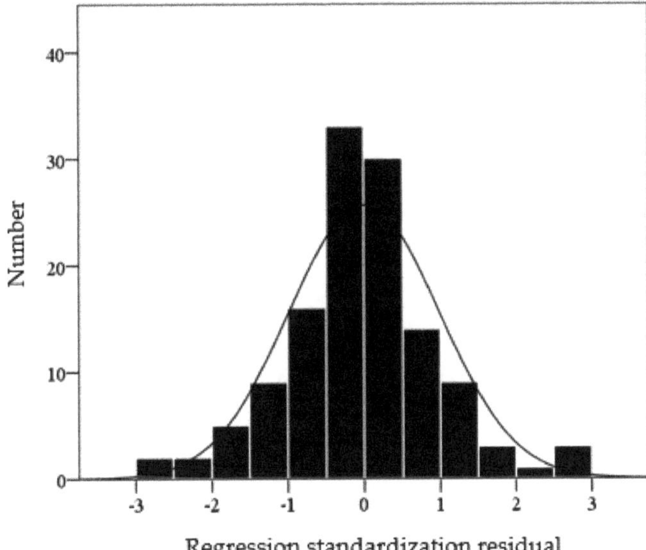

Figure 9. Residual distribution histogram.

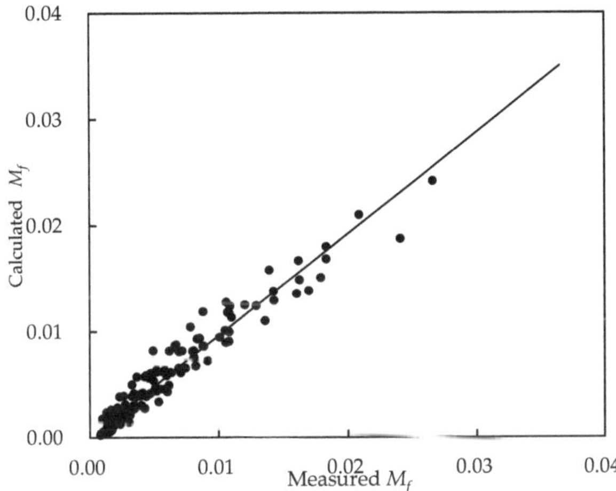

Figure 10. Comparison between the measured and calculated values of M_f.

5. Conclusions

Using a physical model test and theoretical analysis, the factors affecting the hydrodynamic moment M_f of a floating structure were obtained based on dimensional analysis. The characteristics of the moment and the correlations between the influencing factors and the moment were analyzed, and an expression of the moment was derived. The main conclusions are as follows.

The hydrodynamic moment increases with increases in the shape parameter, relative draft depth, relative water level difference and Froude number of a floating structure. The surface pressure difference of a larger structure caused by the uneven distribution of the dynamic pressure and the distance between the action point and the center of the floating structure also increases, thereby increasing the hydrodynamic moment. When the relative water level difference rises, the pressure difference between the front and back surfaces of the structure increases, and the kinetic energy acting on the upstream surface leads to a dynamic pressure increase with a larger Froude number, raising the hydrodynamic moment. Moreover, there are positive linear correlations between the hydrodynamic moment and the shape parameter, relative water level difference and Froude number; however, the relationship between the moment and the ratio of the draft depth to the water level difference is logarithmic.

Based on the results of multivariate linear least squares fitting, stepwise regression analysis was performed to quantitatively obtain a mathematical expression describing the hydrodynamic moment of the floating structure ($1 \leq B/a \leq 4$; $a > h > 0$; $0 < \Delta H/H' \leq 0.10$ and $\mathrm{Fr} \leq 0.35$). Statistical indices were used to quantitatively investigate the accuracy of the formula; the AMCC and SEE values were 0.94 and 0.01, respectively, and the residuals of the formula presented a normal distribution. The obtained formula is easy to calculate and coincides with experimentally measured values and can therefore provide guidance for calculating the hydrodynamic moment of a floating structure.

Author Contributions: Z.C. performed the experiments, analyzed the data and wrote the paper. Y.-J.C. and S.-Y.P. performed the experiments and improved the quality of the paper. The authors also thank the reviewers for their valuable comments, which significantly improved the quality of the paper. All authors have read and agreed to the published version of the manuscript.

Funding: This research was funded by the Changzhou Institute of Technology High-level Talent Research Start-up Funds, grant number (YN 20068).

Institutional Review Board Statement: Not applicable.

Informed Consent Statement: Not applicable.

Data Availability Statement: Not applicable.

Conflicts of Interest: The authors declare no conflict of interest.

Notations

The following symbols are used in this paper:

a	Height of the structure
B	Width of the structure
Fr^2	Froude number: $\mathrm{Fr}^2 = v^2/gH$
g	Gravitational constant of acceleration
H	Upstream water depth
H'	Downstream water depth
h	Draft of the structure
L	Length of the structure
q	Discharge per width of flow
R^2	Coefficient of determination
v	Average flow velocity
ΔH	Water level difference
ρ	Water density
AMCC	Adjusted multiple correlation coefficient
SEE	Standard error of estimation

References

1. Zhang, Q.; Xie, L.H.; Zhou, J. Design of floating gate for overhaul of navigable drainage sluice in the old port of Jin-qing. *Water Resour. Plan. Des.* **2013**, *6*, 80–82.
2. Olyaie, E.; Banejad, H.; Chau, K.; Melesse, A.M. A comparison of various artificial intelligence approaches performance for estimating suspended sediment load of river systems: A case study in United States. *Environ. Monit. Assess.* **2015**, *187*, 189. [CrossRef]
3. Pu, J.H.; Pandey, M.; Hanmaiahgari, P.R. Analytical modelling of sidewall turbulence effect on streamwise velocity profile using 2D approach: A comparison of rectangular and trapezoidal open channel flows. *J. Hydro-Environ. Res.* **2020**, *32*, 17–25. [CrossRef]

4. Pu, J.H. Turbulent rectangular compound open channel flow study using multi-zonal approach. *Environ. Fluid Mech.* **2018**, *19*, 785–800. [CrossRef]
5. Lee, S.; Hong, C. Characteristics of wave exciting forces on a very large floating structure with submerged-plate. *J. Mech. Sci. Technol.* **2005**, *11*, 2061–2067. [CrossRef]
6. Venugopal, V.; Varyani, K.S.; Barltrop, N.D.P. Wave force coefficients for horizontally submerged rectangular cylinders. *Ocean Eng.* **2006**, *33*, 1669–1704. [CrossRef]
7. Roy, P.D.; Ghosh, S. Wave force on vertically submerged circular thin plate in shallow water. *Ocean Eng.* **2006**, *33*, 1935–1953. [CrossRef]
8. Hayatdavoodi, M.; Seiffert, B.; Ertekin, R.C. Experiments and computations of solitary-wave forces on a coastal-bridge deck. Part II: Deck with girders. *Coast. Eng.* **2014**, *88*, 210–228. [CrossRef]
9. Seiffert, B.; Hayatdavoodi, M.; Ertekin, R.C. Experiments and computations of solitary-wave forces on a coastal-bridge deck. Part I: Flat plate. *Coast. Eng.* **2014**, *88*, 194–209. [CrossRef]
10. Xing, D.L. Research on hydrodynamic performance of floating bodies in confined zone. *J. Dalian Univ. Technol.* **1993**, *3*, 351–355.
11. Xing, D.L.; Deng, Y.P.; Zhou, M. Experimental research for added mass of cylinders with reflected boundary condition. *J. Dalian Univ. Technol.* **1998**, *4*, 107–111.
12. Wang, S.; Fu, Z.F.; Cui, Z.; Chen, Y.J. Analysis of the influencing factors of overturning moment on the floating structure surrounded with flowing water. *J. Hohai Univ.* **2018**, *46*, 66–71. [CrossRef]
13. Cui, Z.; Wang, S.; Fu, Z.F. Influence of mesh scale on hydrodynamic characteristics of floating body. *Water Resour. Power* **2016**, *12*, 101–105.
14. Fu, Z.F.; Yin, X.; Gu, X.F. Hydraulic characteristics of floating sluices subsiding and buoying in flowing water. *Adv. Water Resour.* **2014**, *5*, 24–27. [CrossRef]
15. Lu, X.; Wang, K. Modeling a solitary wave interaction with a fixed floating body using an integrated analytical–numerical approach. *Ocean Eng.* **2015**, *109*, 691–704. [CrossRef]
16. Rodrigues, J.M.; Guedes, S.C. A generalized adaptive mesh pressure integration technique applied to progressive flooding of floating bodies in still water. *Ocean Eng.* **2015**, *110*, 140–151. [CrossRef]
17. Cui, Z.; Fu, Z.F.; Chen, Y.J. Explore the flow characteristics of floating structure based on the orthogonal design method. *Eng. J. Wuhan Univ.* **2018**, *6*, 471–477. [CrossRef]
18. Cui, Z.; Fu, Z.F.; Wang, J. Sensitivity analysis of influencing factors of hydraulic characteristics of floating structure in finite flowing water. *Eng. J. Wuhan Univ.* **2018**, *11*, 957–962. [CrossRef]
19. Fu, Z.F.; Yan, Z.M. Stability analysis of a new floating body brake. *J. Hydraulics. Eng.* **2005**, *8*, 1014–1018. [CrossRef]
20. Venugopal, V.; Varyani, K.S.; Westlake, P.C. Drag and inertia coefficients for horizontally submerged rectangular cylinders in waves and currents. *Proc. Inst. Mech. Eng. Part J-J. Eng. Tribol.* **2009**, *223*, 121–136. [CrossRef]
21. Rey, V.; Touboul, J.; Sous, D. Effect of a Submerged Plate on the Near-bed Dynamics under Incoming Waves in Deep Water Conditions. *Appl. Ocean Res.* **2015**, *53*, 67–74. [CrossRef]
22. Johnson, H.K.; Karambas, T.V.; Avgeris, I. Modeling of Waves and Currents around Submerged Breakwaters. *Coast. Eng.* **2005**, *52*, 949–969. [CrossRef]
23. Wan, Z.N.; Zhai, G.G.; Cheng, Y. Prediction on hydro-elastic responses of very large floating structure on the sea. *J. Harbin. Eng. Univ.* **2014**, *10*, 1189–1194. [CrossRef]
24. Zhang, L.P. *Study on Hydraulic Characteristics of Floating Sluice Localization Process*; Hohai University: Nanjing, China, 2007.
25. Rey, V.; Touboul, J. Forces and moment on a horizontal plate due to regular and irregular waves in the presence of current. *Appl. Ocean Res.* **2011**, *33*, 88–99. [CrossRef]
26. Pu, J.H. Velocity profile and turbulence structure measurement corrections for sediment transport-induced water-worked bed. *Fluids* **2021**, *6*, 86. [CrossRef]
27. Pu, J.H.; Wei, J.; Huang, Y.F. Velocity distribution and 3D turbulence characteristic analysis for flow over water-worked rough bed. *Water* **2017**, *9*, 668. [CrossRef]
28. Li, Y.; Lin, M. Hydrodynamic coefficients induced by waves and currents for submerged circular cylinder. *Procedia Eng.* **2010**, *4*, 253–261. [CrossRef]
29. Cui, Z.; Fu, Z.F.; Dai, W.H.; Lai, Z.Q. Submerged fixed floating structure under the action of surface current. *Water* **2018**, *10*, 102. [CrossRef]
30. Goring, D.G.; Nikora, V.I. Despiking acoustic doppler velocimeter data. *J. Hydraul. Eng.* **2002**, *128*, 117–126. [CrossRef]
31. Blanckaert, K.; Lemmin, U. Means of noise reduction in acoustic turbulence measurements. *J. Hydraul. Res.* **2006**, *44*, 1–15. [CrossRef]

MDPI
St. Alban-Anlage 66
4052 Basel
Switzerland
Tel. +41 61 683 77 34
Fax +41 61 302 89 18
www.mdpi.com

Fluids Editorial Office
E-mail: fluids@mdpi.com
www.mdpi.com/journal/fluids

www.ingramcontent.com/pod-product-compliance
Lightning Source LLC
LaVergne TN
LVHW070637170726
843515LV00004B/271
* 9 7 8 3 0 3 6 5 3 2 4 2 4 *